BestMasters

Mit „**BestMasters**" zeichnet Springer die besten Masterarbeiten aus, die an renommierten Hochschulen in Deutschland, Österreich und der Schweiz entstanden sind. Die mit Höchstnote ausgezeichneten Arbeiten wurden durch Gutachter zur Veröffentlichung empfohlen und behandeln aktuelle Themen aus unterschiedlichen Fachgebieten der Naturwissenschaften, Psychologie, Sozialwissenschaften, Technik und Wirtschaftswissenschaften. Die Reihe wendet sich an Praktiker und Wissenschaftler gleichermaßen und soll insbesondere auch Nachwuchswissenschaftlern Orientierung geben.

Springer awards **"BestMasters"** to the best master's theses which have been completed at renowned Universities in Germany, Austria, and Switzerland. The studies received highest marks and were recommended for publication by supervisors. They address current issues from various fields of research in natural sciences, psychology, social sciences, technology, and economics. The series addresses practitioners as well as scientists and, in particular, offers guidance for early stage researchers.

Nikolas Uesseler

Parameter Identification for a Stochastic Partial Differential Equation in the Nonstationary Case

Nikolas Uesseler (ORCID)
Fachbereich Mathematik
Universität Münster
Münster, Germany

ISSN 2625-3577   ISSN 2625-3615 (electronic)
BestMasters
ISBN 978-3-658-50343-7   ISBN 978-3-658-50344-4 (eBook)
https://doi.org/10.1007/978-3-658-50344-4

This Springer Spektrum imprint is published by the registered company Springer Fachmedien Wiesbaden GmbH, part of Springer Nature.
The registered company address is: Abraham-Lincoln-Str. 46, 65189 Wiesbaden, Germany

If disposing of this product, please recycle the paper.

Competing Interests The author has no competing interests to declare that are relevant to the content of this manuscript.

Introduction

The final goal of this thesis is to analyze a concrete ill-posed problem in the world of Parameter Identification of Partial Differential Equations. We want to study the existing results on the topic in order to apply them later to a problem arising from research in biology.

In the first section we will revisit the basic theory of nonlinear ill-posed problems. This is done to show the essential ideas of Tikhonov regularization, which is important to understand the generalizations that came in the last years. In the second section we focus on the topic of the convergence speed in the regularization process. Different assumptions on the ground truth are possible in order to get this kind of results. In the next section we go through the generalizations of the regularization process that made possible to work with a bigger family of ill-posed problems. For example, problems in which the measured data is of random nature. In fact, for this kind of problems some tools from probability theory are needed, which are the topic of section four. We finish with two applications in the last section motivated by an interesting problem in biology. Both applications are based on the relation between Stochastic Differential Equations and the Fokker–Planck equation. The first application assumes that the underlying probability distribution is stationary and the second assumption generalizes the ideas to the nonstationary case. This work is the basis of the article [25], in which the results can be seen in a more compact version.

The path that we will take to gain understanding of the theoretical frame will be the study of its historical development. This is done with the conviction that great part of understanding an idea is understanding the history and evolution of the idea.

Contents

Tikhonov regularization in Nonlinear Problems

1

A good place to start our historical study could be the publication in which the theory of ill-posed problems that was already known to that time was extended to the case of nonlinear operators. T.Seidman and C.Vogel published in 1988 the paper *Well posedness and convergence of some regularization methods for non-linear ill posed problems* [24]. The setting they consider is the following:

Let $\hat{F}(x) = \hat{y}$ be a nonlinear problem, where $\hat{F} : X \to Y$ is an operator between Banach spaces. The ill-posedness here means that perturbing the data or the operator delivers a problem $F(x) = y$ which does not have to posses a solution and if it has a solution it will not necessarily depend continuously on small changes of F or y. In this setting the choice of the modeling operator F is understood together with y as the data. The strategy of regularization consists in finding a solution x_μ of the problem that corresponds to the measured data $[F, y]$ together with a parameter μ of the scheme we choose to approximate the original problem $\hat{F}(x) = \hat{y}$. We hope for x_μ to be in fact an approximation of the exact solution x. There are three properties that we would like to have in our scheme:

1. Existence: a triple $[F, y, \mu]$ should deliver a (hopefully unique) solution x_μ
2. Stability: for a fixed μ, x_μ should depend continuously on the data $[F, y]$ and
3. Convergence: for a sequence $[F, y]_k \to [F, y]$, there should be an appropriate choice of parameters μ_k, such that $x_{\mu_k} \to x$.

The method of regularization that is considered here is the *Tikhonov regularization*. Namely, the abstract regularization parameter μ is now represented by the positive number α and one defines an approximating solution x_α as the minimizer of the Tikhonov functional

N. Uesseler, *Parameter Identification for a Stochastic Partial Differential Equation in the Nonstationary Case*, BestMasters,
https://doi.org/10.1007/978-3-658-50344-4_1

1

$$T_\alpha^{y^\delta}(x) := \|F(x) - y^\delta\|_Y^2 + \alpha J(x) \ \textit{for} \ x \in X,$$

where $J : X \to \mathrm{R}$ is a penalty function that incorporates a-priori knowledge. We can see that this setting is of great generality. It is based on Banach spaces and the penalty function is not specified. We will come back later to the conditions that J would have to fulfill.

A couple of months later in 1989 a result on convergence rates was published by Engl, Kunisch and Neubauer in *Convergence rates for Tikhonov regularization of nonlinear ill-posed problems* [9]. Although convergence of the approximated solution to the "true" solution is of obvious theoretical importance, in applications it is crucial to know as well the rate of convergence, since in general the convergence might be arbitrarily slow. Thus we get a fourth topic of study:

4. Convergence rates: estimates of the error of the approximated solution depending on the modeling and measurement error.

The basic results regarding these four topics were presented very clearly in the book *regularization of inverse problems* by Engl, Hanke and Neubauer, published in 1996 [8]. They focus on the special case of Hilbert spaces, penalty function defined by $J(x) = \|x - x^\star\|^2$ for some fixed $x^\star$ and exact operator $F = \hat{F}$. In order to have these basic and important results at hand together with their proofs we will go shortly through this presentation. However the main goal of the section will be identifying the essential properties of the objects that are used in the proofs in order to comfortably understand the generalizations that came later.

In this section about the Tikhonov regularization we will then consider the following setting: $F(x) = y$ is a nonlinear problem for which an exact solution exists and $F : X \to Y$ is an operator between Banach spaces with the following properties:

- F is continuous
- F is weakly sequentially continuous and
- $\mathcal{D}(F)$ is weakly closed.

We will comment as well on the special results we would get if instead of Banach spaces we considered Hilbert-spaces.

Let us first prove that under these assumptions, a so called $x^\star-minimum\ norm$ *solution* x exists. this is, x minimizes $\|x - x^\star\|^2$ over the set of all solutions of $F(x) = y$.

Proposition 1.1 *Let $F : X \to Y$ be an operator between reflexive Banach spaces with weakly closed Domain and which is continuous and weakly continuous. If an exact solution of $F(x) = y$ exists, then the problem has an $x^\star-$minimum norm solution.*

Proof Consider the nonempty set of solutions $S := \{x \in X : F(x) = y\}$. First we will see that this set is weakly closed due to the weak continuity of F: let $\{x_n\} \subset S$ such that $x_n \rightharpoonup x$. Then $F(x_n) = y$ for all n and thus $F(x_n) \to y$. By weak continuity we get $F(x) = y$, which means $x \in S$.

Now let us take a sequence $\{x_n\} \subset S$, such that $\|x_n - x^\star\| \to \inf_{x \in S} \|x - x^\star\| \geq 0$ monotonically. Then we have $\|x_n - x^\star\| \leq \|x_0 - x^\star\| < \infty$, which implies that x_n is bounded. Thus we can use the Banach–Alaouglu theorem to find a weakly convergent subsequence $x_n \rightharpoonup x$. As we saw, x is in S as a weakly closed set, which means that x is a solution. To see that x is also an $x^\star-$ minimum norm solution we use the lower semi-continuity of the norm. Namely

$$\|x - x^\star\| \leq \liminf_{n \to \infty} \|x_n - x^\star\| = \inf_{x \in S} \|x - x^\star\|.$$

Thus x attains the infimum and is an $x^\star-$minimum norm solution. $\square$

Remark 1.1 *Notice, since it will be important later, that the existence of the $x^\star-$ minimum norm solution is not based on all the defining properties of the norm, but only on two facts that are essential. The first one is the weak lower semi continuity of the norm and the second one is that we can use Banach–Alaouglu. In fact our proof would have been analogous if instead of the norm, we would have used a weakly lower semi-continuous functional whose sublevelsets were compact with respect to the weak topology and nonempty for nonnegativ levels. We could even go a step further and replace the weak topology with some topology τ, as long as the functional still has the two mentioned properties. Then we would replace the weak continuity of F by continuity with respect to τ.*

1.1   Ill-posedness Condition

First we have a look at an interesting result regarding the question whether some given problem is actually ill-posed or not.

Proposition 1.2 *Let F be a nonlinear, compact and weakly closed operator. Assume $F(x^\dagger) = y$ and that there is an $\epsilon > 0$ such that $F(x) = \hat{y}$ has a unique solution for all $\hat{y} \in B_\epsilon(y) \cap \mathcal{R}(F)$. If there is a sequence $\{x_n\} \subset \mathcal{D}(F)$, such that*

$$x_n \rightharpoonup x^\dagger \text{ but } x_n \nrightarrow x^\dagger, \tag{1.1}$$

then F^{-1} defined on $B_\epsilon(y) \cap \mathcal{R}(F)$ is not continuous.

Proof Let us take such a sequence $\{x_n\} \subset \mathcal{D}(F)$ with $x_n \rightharpoonup x^\dagger$ and $x_n \nrightarrow x^\dagger$. Since $\mathcal{D}(F)$ is weakly closed $x^\dagger$ is also in $\mathcal{D}(F)$. Using the facts that weakly convergent sequences are bounded and that F is compact, we find a subsequence $\{x_n\}$ (without changing the notation) such that $F(x_n) \to \hat{y}$ and especially $F(x_n) \rightharpoonup \hat{y}$. By weak continuity of F this implies $F(x^\dagger) = \hat{y}$ and thus we see strong convergence $F(x_n) \to F(x^\dagger)$.

Now for large n, $F(x_n) \in B_\epsilon(y) \cap \mathcal{R}(F)$ and x_n is the unique solution of the problem $F(x) = F(x_n) =: y_n$, i.e. $x_n = F^{-1}(y_n)$. Thus we have

$$y_n \to \hat{y} \text{ but by assumption } F^{-1}(y_n) = x_n \nrightarrow x^\dagger = F^{-1}(\hat{y}), \tag{1.2}$$

which means that F^{-1} is discontinuous. $\square$

Notice that the condition of uniqueness close to y can be understood as injectivity around $x^\dagger$ and that the existence of the sequence that we used is only possible if $\mathcal{D}(F)$ is infinite dimensional around $x^\dagger$, otherwise weakly convergent sequences converge also in the strong sense.

1.2   Existence of the Minimizer

We already proved the existence of an $x^\star-$ minimum norm solution. Now we want to approximate it by minimizing the Tikhnov functional $T_\alpha^{y^\delta}$ defined as above, where y^δ is a measurement which we assume approximates the exact data y. However we need to check if the minimizer of this functional actually exists. The proof is very similar to the proof of the existence of the $x^\star-$ minimum norm solution.

Theorem 1.1 *Let $F : X \to Y$ be an operator between reflexive Banach spaces with weakly closed domain and which is continuous and weakly continuous. Let us also assume, as before that an exact solution of $F(x) = y$ exists. Then the functional*

$$T_\alpha^{y^\delta}(x) := \| F(x) - y^\delta \|_Y^2 + \alpha \| x - x^\star \|^2 \ \text{for } x \in X,$$

posseses a minimizer $x_\alpha^{y^\delta}$ *(also denoted* T_α^δ *and* x_α^δ*).*

Proof First see that the functional is not everywhere ∞ since a solution of $F(x) = y$ is assumed to exist and thus it is in $\mathcal{D}(F)$, on this point the functional takes a finite value.

Let us now consider a sequence $\{x_n\} \subset \mathcal{D}(F)$, such that

$$T_\alpha^\delta(x_n) \ \to \ \inf_{x \in \mathcal{D}(F)} T_\alpha^\delta(x) \geq 0$$

monotonically. Then it holds

$$\alpha \| x_n - x^\star \|^2 \leq T_\alpha^\delta(x_n) \leq T_\alpha^\delta(x_0) < \infty,$$

as well as

$$\| F(x_n) - y^\delta \|^2 \leq T_\alpha^\delta(x_n) \leq T_\alpha^\delta(x_0) < \infty.$$

Now using in both spaces X and Y the theorem of Banach–Alaouglu and the assumption that $\mathcal{D}(F)$ is weakly closed we get weakly convergent subsequences (without renaming them) with

$$x_n \rightharpoonup \hat{x} \in \mathcal{D}(F) \ \text{ and } \ F(x_n) \rightharpoonup \hat{y}.$$

By weak continuity of F it hold as well $F(\hat{x}) = \hat{y}$. Now we prove that the infimum of T_α^δ is in fact attained by this weak limit $\hat{x}$. Namely by weak lower semi-continuity of the norm used twice we see

$$\| \hat{x} - x^\star \| \leq \liminf_{n \to \infty} \| x_n - x^\star \|$$

and

$$\| F(\hat{x}) - y^\delta \| \leq \liminf_{n \to \infty} \| F(x_n) - y^\delta \|.$$

Finally adding both inequalities we see that $\hat{x}$ is a minimizer:

$$T_\alpha^\delta(\hat{x}) \leq \liminf_{n \to \infty} T_\alpha^\delta(x_n) \leq \inf_{x \in \mathcal{D}(F)} T_\alpha^\delta(x).$$

$\square$

Remark 1.2 *Notice that for this argument the essential properties of the penalty term are the same as the ones we mentioned in remark 1.1. Let us now also comment on the essential properties of the so-called fidelity term $\|F(x) - y^\delta\|^2$ that were used in the proof. Later this term will be generalized by some functional $S(F(x), y^\delta)$, so we should keep track of the properties that we need. As for the penalty term (again by remark 1.1), the Banach–Alaouglu theorem is not essential if the sublevelsets of $S(., y^\delta)$ are compact with respect to some weaker topology that is of our interest. Moreover, this time we used also for this term lower semi-continuity.*

1.3   Stability of the Regularization

By the last subsection we can define now a map $y^\delta \mapsto x_\alpha^\delta$, where x_α^δ is a minimizer of the Tikhonov functional T_α^δ. Obviously we want this map to be continuous. First we will look at continuity referring to the weak topology on X and strong topology on Y. In the special case of Hilbert spaces, the continuity with respect to the strong topology follows as a corollary. Let us prove these facts.

Theorem 1.2 *Let $F : X \to Y$ be an operator between Banach spaces fulfilling the assumptions needed for the existence of an $x^* -$ minimum norm solution and for the existence of the minimizer of the Tikhonov functional T_α^δ. Let $y_n \to y^\delta$ in Y and let x_n be a minimizer of $T_\alpha^{y_n}$. Then x_n has a weakly convergent subsequence and every weak limit is a minimizer of T_α^δ.*

Proof By the minimizing property of x_n and by the continuity of the norm w.r.t. the strong topology we have

$$T_\alpha^{y_n}(x_n) \le T_\alpha^{y_n}(x^\star) = \|F(x^\star) - y_n\|^2 \to \|F(x^\star) - y^\delta\| < \infty$$

for all n that are large enough. Thus the terms

$$\|F(x_n) - y_n\|^2 \text{ and } \|x^\star - x_n\|^2$$

in the Tikhonov functional $T_\alpha^{y_n}(x_n)$ are bounded. This implies again with continuity of the norm using the subsequence that defines the limes superior

$$\limsup_{n\to\infty} \|F(x_n) - y^\delta\| = \limsup_{n\to\infty} \|F(x_n) - y_n\| < \infty,$$

which means that $\|F(x_n) - y^\delta\|$ is also bounded. Hence we can apply as usual Banach–Alaouglu on the sequences $\{x_n\}$ and $\{F(x_n)\}$ to find convergent subsequences

$$F(x_n) \rightharpoonup \hat{y} \text{ and } x_n \rightharpoonup \hat{x}$$

and by weak convergence $F(\hat{x}) = \hat{y}$. Now we prove that this weak limit is a minimizer of T_α^δ. First we make use of weak lower semi-continuity in both terms of the Tikhonov functional to get

$$\|\hat{x} - x^\star\| \le \liminf_{n \to \infty} \|x_n - x^\star\|$$

and

$$\|F(\hat{x}) - y^\delta\| \le \liminf_{n \to \infty} \|F(x_n) - y_n\|.$$

Then we add these inequalities and use the minimizing property of x_n, as well as the strong convergence $y_n \to y^\delta$ together with the continuity of the norm. For any $z \in \mathcal{D}(F)$ we obtain:

$$
\begin{aligned}
T_\alpha^\delta(\hat{x}) &\le \liminf_{n \to \infty} T_\alpha^{y_n}(x_n) \\
&\le \limsup_{n \to \infty} T_\alpha^{y_n}(x_n) \\
&\le \limsup_{n \to \infty} T_\alpha^{y_n}(z) \\
&= \limsup_{n \to \infty} \|F(z) - y_n\|^2 + \alpha\|z - x^\star\|^2 \\
&= \|F(z) - y^\delta\|^2 + \alpha\|z - x^\star\|^2 \\
&= T_\alpha^\delta(z).
\end{aligned}
$$

$\square$

Before we prove that in the special case of Hilbert spaces the continuity is even with respect to the strong topology on X, we make a remark on the essential properties needed in the last proof.

Remark 1.3 *In this proof we did not make use of any other property of the penalty term we have not already highlighted in remark 1.1. In the case of the fidelity term however, the continuity of the norm $\|.\|_Y$ (which is contained in the fidelity term) w.r.t the strong topology appeared in an essential way. More specifically, we used the fact that $v_n \to v \implies \|v' - v_n\| \to \|v' - v\|$. Later we will replace the fidelity term and the strong topology, but these properties are necessary for the result.*

Theorem 1.3 *Let $F : X \to Y$ fulfill the same assumptions as in the last theorem but be defined between two Hilbert spaces. Let as before $y_n \to y^\delta$ in Y and let x_n be a minimizer of $T_\alpha^{y_n}$. Then x_n has a convergent subsequence and every limit is a minimizer of T_α^δ, where the convergence is with respect to the strong topology.*

Proof From the last chain of inequalities in the proof of the theorem above, setting $z = \hat{x}$, we obtain

$$T_\alpha^\delta(\hat{x}) \leq \liminf_{n\to\infty} T_\alpha^{y_n}(x_n) \leq \limsup_{n\to\infty} T_\alpha^{y_n}(x_n) \leq T_\alpha^\delta(\hat{x})$$

and thus

$$\lim_{n\to\infty} T_\alpha^{y_n}(x_n) = T_\alpha^\delta(\hat{x}), \tag{1.3}$$

where $x_n \rightharpoonup \hat{x}$ was a weakly convergent subsequence of the sequence of minimizers we defined with $F(x_n) \rightharpoonup F(\hat{x})$. It will follow that the convergence is indeed strong. Assume by contradiction that $x_n \not\to \hat{x}$. Then it holds also $\|x_n - x^\star\| \not\to \|\hat{x} - x^\star\|$, otherwise this together with the weak convergence $x_n - x^\star \rightharpoonup \hat{x} - x^\star$ would imply already strong convergence in a Hilbert space. Since the weak lower semi-continuity

$$\|\hat{x} - x^\star\| \leq \liminf_{n\to\infty} \|x_n - x^\star\|$$

is given, upper semi continuity

$$\limsup_{n\to\infty} \|x_n - x^\star\| \leq \|\hat{x} - x^\star\|$$

has to fail. From last proof we know x_n was bounded and thus

$$\|\hat{x} - x^\star\| < \limsup_{n\to\infty} \|x_n - x^\star\| =: C < \infty$$

By taking a proper subsequence we can say

$$x_n \rightharpoonup \hat{x}, \quad F(x_n) \rightharpoonup F(\hat{x})$$

and now also $\|x_n - x^\star\| \to C$.

By rearranging the terms of equality (1.3) we obtain

$$\lim_{n\to\infty} \|F(x_n) - y_n\|^2 = \|F(\hat{x}) - y^\delta\|^2 + \alpha(\|\hat{x} - x^\star\|^2 - C^2)$$
$$< \|F(\hat{x}) - y^\delta\|^2$$

But this directly contradicts the weak lower semi-continuity

$$\|F(\hat{x}) - y^\delta\| \leq \liminf_{n\to\infty} \|F(x_n) - y_n\|.$$

This proves that $x_n \to \hat{x}$ in the strong topology. $\qquad\qquad\square$

1.4 Convergence of the Regularization

The map $y^\delta \mapsto x_\alpha^\delta$ is now well-defined and continuous in y^δ with respect to the weak topology on X and strong topology on Y. The next question to adress is if the approximate solution x_α^δ converges to an $x^\star-$ minimum norm solution $x^\dagger$ as we decrease the penalization parameter, namely as $\alpha \to 0$. Since the problem is ill-posed, we do not expect this to happen unless we improve **significantly** the quality of our measurements y^δ as well. Or conversely, for measurements $y^\delta \to y$ that get progressively better we are allowed to improve **moderately** the penalization parameter α and thus approximate more accurately the original problem $F(x) = y$. The next theorem quantifies the relation between the error of the measurements and the penalization parameter that is necessary for the approximate solution x_α^δ to converge to the exact solution $x^\dagger$. From now on we will assume $\|y^\delta - y\| < \delta$, which quantifies the error of the measurement and justifies the notation y^δ we have been using.

Theorem 1.4 *Let $F : X \to Y$ be an operator between Banach spaces fulfilling the assumptions needed for the existence of an $x^\star-$ minimum norm solution and for the existence of a minimizer x_α^δ of the Tikhonov functional T_α^δ. Let $y^\delta \in Y$ be a measurement such that $\|y^\delta - y\| \leq \delta$ and let the penalization parameter $\alpha(\delta)$ be such that $\alpha(\delta) \to 0$ and $\delta^2/\alpha(\delta) \to 0$ as $\delta \to 0$. Then every sequence $\{x_\alpha^\delta\}$ has* *a weakly convergent subsequence and every weak limit is an $x^\star-$ minimum norm solution.*

Proof Let $x^\dagger$ be an $x^\star-$minimum norm solution. By the minimizing property of x_α^δ we have

$$T_\alpha^\delta(x_\alpha^\delta) \le T_\alpha^\delta(x^\dagger) = \|F(x^\dagger) - y^\delta\|^2 + \alpha\|x^\star - x^\dagger\|^2 \le \delta^2 + \alpha\|x^\star - x^\dagger\|^2$$

for all α small enough. In particular both terms in $T_\alpha^\delta(x_\alpha^\delta)$ fulfill

$$\alpha\|x^\star - x_\alpha^\delta\|^2 \le \delta^2 + \alpha\|x^\star - x^\dagger\|^2 \quad \text{and}$$
$$\|F(x_\alpha^\delta) - y^\delta\|^2 \le \delta^2 + \alpha\|x^\star - x^\dagger\|^2$$

and thus

$$\|x^\star - x_\alpha^\delta\|^2 \le \delta^2/\alpha + \|x^\star - x^\dagger\|^2. \tag{1.4}$$

By the condition on δ^2/α, $\{x_\alpha^\delta\}$ is bounded and there is a weakly convergent subsequence $x_\alpha^\delta \rightharpoonup \hat{x}$. Moreover it holds by continuity of the norm

$$\begin{aligned}
\limsup_{\delta \to 0} \|F(x_\alpha^\delta) - y\| &= \limsup_{\delta \to 0} \|F(x_\alpha^\delta) - y^\delta\| \\
&\le \limsup_{\delta \to 0} \sqrt{\delta^2 + \alpha}\|x^\star - x^\dagger\| \\
&= 0.
\end{aligned}$$

By the weak continuity of F we know $F(\hat{x}) = y$ and hence $\hat{x}$ is a solution. To see that it is also an $x^\star-$minimum norm solution we use the weak lower semicontinuity and inequality (1.4):

$$\begin{aligned}
\|x^\star - \hat{x}\| &\le \liminf_{\delta \to 0} \|x^\star - x_\alpha^\delta\| \\
&\le \limsup_{\delta \to 0} \|x^\star - x_\alpha^\delta\| \\
&\le \|x^\star - x^\dagger\| \\
&\le \|x^\star - \hat{x}\|,
\end{aligned}$$

where the last inequality is the minimality of $x^\dagger$. In particular it holds $\|x^\star - x^\dagger\| = \|x^\star - \hat{x}\|$ and thus $\hat{x}$ is indeed an $x^\star-$minimum norm solution. $\qquad\square$

Remark 1.4 *We did not use any other property of the fidelity or the penalty term that we have not already discussed in the remarks 1.1, 1.2 and 1.3 to prove last theorem.*

As in the case of stability, if $F : X \to Y$ is defined between Hilbertspaces, then the claim holds for the strong norm-topolgy on X as well.

Corollary 1.1 *Under the same assumptions and notation of last theorem, if $F : X \to Y$ is defined between Hilbert spaces, the convergence $x_\alpha^\delta \rightharpoonup \hat{x}$ holds also in the strong topology, namely $x_\alpha^\delta \to \hat{x}$*

Proof Using the the last chain of inequalities in the proof of last theorem and the weak convergence $x_\alpha^\delta \rightharpoonup \hat{x}$, we get:

$$\limsup_{\delta \to 0} \|x_\alpha^\delta - \hat{x}\|^2 = \underbrace{\limsup_{\delta \to 0} \|x_\alpha^\delta - x^\star\|^2 + \|x^\star - \hat{x}\|^2}_{\leq \|x^\star - \hat{x}\|^2} - \underbrace{2\limsup_{\delta \to 0} \langle x_\alpha^\delta - x^\star, \hat{x} - x^\star \rangle}_{= \|\hat{x} - x^\star\|^2}$$

$$\leq 0.$$

And thus we have strong convergence in this situation.

1.5 Convergence Rates

To finish this section we will look at the so-called convergence rates in Hilbert spaces. We know that the Tikhonov regularized solution x_α^δ exists, is stable and for a correctly chosen parametrisation it converges to an $x^\star-$minimum norm solution $x^\dagger$. This convergence can be in general arbitrarily slow unless we introduce more a-priori knowledge on the ground truth $x^\dagger$. This knowledge can be expressed as a so-called source condition. Let us try to motivate this concept first.

As before we assume that the ground truth $x^\dagger$ is an $x^\star-$minimum norm solution, namely a solution of the problem

$$minimize \quad \frac{1}{2}\|x^\star - x\|^2 \quad subject\ to \quad F(x) = y.$$

Naturally we can consider the Lagrangian

$$L(\omega, x) = \frac{1}{2}\|x^\star - x\|^2 + (\omega, F(x) - y), \qquad (1.5)$$

where $\omega \in Y^\star$ and $(.,.)$ the respective dual pairing. Let us assume that F is Fréchet-differentiable and that strong duality holds, this is, if $x^\dagger \in X$ is the minimizer of the problem, there exists an $\hat{\omega} \in Y^\star$ such that $(\hat{\omega}, x^\dagger)$ is a saddle point of L. In particular, under this assumption it holds

$$\frac{\partial}{\partial x} L(\hat{\omega}, x)|_{x^\dagger} = 0.$$

which is equivalent to

$$x^\star - x^\dagger = F'(x^\dagger)^\# \omega,$$

understood as elements of $X^\star$. If for instance we considered the linear case in the setting of Hilbert spaces with $x^\star = 0$ we arrive to the well known source condition $x^\dagger = F'(x^\dagger)^\# \omega = (F'(x^\dagger)^\# F'(x^\dagger))^{\frac{1}{2}} p$, which is to be understood as imposed regularity of $x^\dagger$ measured in terms of F. From this point of view this regularity assumption can be generalized by the first assumption we made, namely that strong duality holds. And in fact, for the general nonlinear problem this assumption allows us to achieve convergence rates results. Let us sum up what we need for the next result to hold.

Assumption 1.1 *X is a Hilbert space, Y is a Hilbert space and $F : X \to Y$ is Fréchet-differentiable with convex domain $\mathcal{D}(F)$.*

$x^\dagger$ is an $x^\star-$minimum norm solution that fulfills a source condition, i.e there exists $\omega \in Y$ satisfying

$$x^\dagger - x^\star = F'(x^\dagger)^\# \omega.$$

F fulfills the following nonlinearity condition: there exists $L > 0$ with

$$\|F'(x^\dagger) - F'(z)\| \le L\|x^\dagger - z\| \text{ for all } z \in X$$

The Lagrange multiplier ω is small enough, namely

$$L\|\omega\| < 1.$$

Theorem 1.5 *Let assumption 1.1 hold and let $y^\delta \in Y$ be measurements of y with $\|y^\delta - y\| \leq \delta$. Let as before x_α^δ be the minimizer of the Tikhonov functional T_α^δ. Then for the choice $\alpha \sim \delta$ it holds*

$$\|x_\alpha^\delta - x^\dagger\| \sim \sqrt{\delta}.$$

Proof By definition of x_α^δ and $x^\dagger$ we know

$$T_\alpha^\delta(x_\alpha^\delta) \leq T_\alpha^\delta(x^\dagger) \leq \delta^2 + \alpha\|x^\dagger - x^\star\|^2.$$

By adding $\alpha\|x_\alpha^\delta - x^\dagger\|^2$ on both sides, rearranging the terms and using the source condition we obtain

$$\begin{aligned}
\|F(x_\alpha^\delta) - y^\delta\|^2 + \alpha\|x_\alpha^\delta - x^\dagger\|^2 &\leq \delta^2 + \alpha(\|x^\dagger - x^\star\|^2 + \|x^\dagger - x_\alpha^\delta\|^2 - \|x_\alpha^\delta - x^\star\|^2) \\
&= \delta^2 + 2\alpha(x^\dagger - x_\alpha^\delta, x^\dagger - x^\star)_X \\
&= \delta^2 + 2\alpha(x^\dagger - x_\alpha^\delta, F'(x^\dagger)^\# \omega)_X \\
&= \delta^2 + 2\alpha(F'(x^\dagger)(x^\dagger - x_\alpha^\delta), \omega)_Y.
\end{aligned}$$

The next step, which involves the assumed Fréchet-differentiability, is to prove the following equality:

$$F(x_\alpha^\delta) = F(x^\dagger) + F'(x^\dagger)(x_\alpha^\delta - x^\dagger) + r_\alpha^\delta,$$

where $\|r_\alpha^\delta\| \leq \frac{1}{2}L\|x_\alpha^\delta - x^\dagger\|^2$. To see that let us define $f(t) := F(x^\dagger + t(x_\alpha^\delta - x^\dagger))$ and thus $f'(t) = F'(x^\dagger + t(x_\alpha^\delta - x^\dagger))(x_\alpha^\delta - x^\dagger)$. Then we see

$$\begin{aligned}
\|f'(t) - f'(\tilde{t})\| &\leq \|F'(x^\dagger + t(x_\alpha^\delta - x^\dagger)) - F'(x^\dagger + \tilde{t}(x_\alpha^\delta - x^\dagger))\|\|x_\alpha^\delta - x^\dagger\| \\
&\leq L\|(\tilde{t} - t)(x_\alpha^\delta - x^\dagger)\|\|x_\alpha^\delta - x^\dagger\| \\
&\leq \underbrace{L\|x_\alpha^\delta - x^\dagger\|^2}_{\text{Lipschitz constant of } f'}(t - \tilde{t}).
\end{aligned}$$

From this calculation it follows

$$
\begin{aligned}
\|r_\alpha^\delta\| &:= \|F(x_\alpha^\delta) - F(x^\dagger) + F'(x^\dagger)(x_\alpha^\delta - x^\dagger)\| \\
&= \|f(1) - f(0) - f'(0)\| \\
&= \left\| \int_0^1 f'(t) - f'(0)\, dt \right\| \\
&\le L\|x_\alpha^\delta - x^\dagger\|^2 \int_0^1 t\, dt \\
&= \frac{L}{2}\|x_\alpha^\delta - x^\dagger\|^2.
\end{aligned}
$$

Combining this with the first estimate we found, we see

$$
\begin{aligned}
\|F(x_\alpha^\delta) - y^\delta\|^2 &+ \alpha\|x_\alpha^\delta - x^\dagger\|^2 \\
&\le \delta^2 + 2\alpha(F'(x^\dagger)(x^\dagger - x_\alpha^\delta), \omega)_Y \\
&= \delta^2 + 2\alpha(F(x_\alpha^\delta) - F(x^\dagger) - r_\alpha^\delta, \omega)_Y \\
&= \delta^2 + 2\alpha(F(x_\alpha^\delta) - y^\delta + y^\delta - F(x^\dagger) - r_\alpha^\delta, \omega)_Y \\
&\le \delta^2 + 2\alpha\|F(x_\alpha^\delta) - y^\delta\|\|\omega\| + 2\alpha\|y^\delta - F(x^\dagger)\|\|\omega\| + 2\alpha\|r_\alpha^\delta\|\|\omega\| \\
&\le \delta^2 + 2\alpha\|F(x_\alpha^\delta) - y^\delta\|\|\omega\| + 2\alpha\delta\|\omega\| + \alpha L\|x_\alpha^\delta - x^\dagger\|^2\|\omega\|.
\end{aligned}
$$

Now by rearranging the terms we get

$$
\|F(x_\alpha^\delta) - y^\delta\|^2 + \alpha(1 - L\|\omega\|)\|x_\alpha^\delta - x^\dagger\|^2 \le \delta^2 + 2\alpha\delta\|\omega\| + 2\alpha\|F(x_\alpha^\delta) - y^\delta\|\|\omega\|.
$$

First we omit the second term in the left using that $(1 - L\|\omega\|) > 0$ and obtain by the general implication

$$
a^2 \le ab + c^2 \implies a \le b + c \quad \text{for } a, b, c \ge 0
$$

that

$$
\underbrace{\|F(x_\alpha^\delta) - y^\delta\|^2}_{=:a^2} \le \underbrace{\delta^2 + 2\alpha\delta\|\omega\|}_{=:c^2} + \underbrace{2\alpha\|\omega\|}_{b}\ \underbrace{\|F(x_\alpha^\delta) - y^\delta\|}_{a}
$$

$$
\implies \|F(x_\alpha^\delta) - y^\delta\| \le (\delta^2 + 2\alpha\delta\|\omega\|)^{\frac{1}{2}} + 2\alpha\|\omega\|.
$$

Now we omit the first term in the left and use that $(1 - L\|\omega\|) > 0$. We finally see

$$
\begin{aligned}
\|x_\alpha^\delta - x^\dagger\|^2 &\leq \frac{\delta^2 + 2\alpha\delta\|\omega\| + 2\alpha\|F(x_\alpha^\delta) - y^\delta\|\|\omega\|}{\alpha(1 - L\|\omega\|)} \\[2mm]
&\leq \frac{\delta^2 + 2\alpha\delta\|\omega\| + 2\alpha\|\omega\|(\delta^2 + 2\alpha\delta\|\omega\|)^{\frac{1}{2}} + (2\alpha\|\omega\|)^2}{\alpha(1 - L\|\omega\|)} \\[2mm]
&= \frac{((\delta^2 + 2\alpha\delta\|\omega\|)^{\frac{1}{2}} + 2\alpha\|\omega\|)^2}{\alpha(1 - L\|\omega\|)}.
\end{aligned}
$$

With the choice $\alpha \sim \delta$ we get indeed

$$
\|x_\alpha^\delta - x^\dagger\| \sim \sqrt{\delta}.
$$

$\square$

We have seen the basic arguments that justify the Tikhonov regularization approach theoretically. We have analyzed **existe**nce of an $x^\star$−minimum norm soltion $x^\dagger$, existence of a regularized solution x_α^δ, stability of the regularized solution with respect to its data y^δ and convergence of x_α^δ to $x^\dagger$ as the measurements y^δ approach the exact data y. All these results are valid in the context of an operator $F : X \to Y$ between two Banach spaces that is sequentially continuous with respect to the weak topologies of X and Y. To be specific, the concepts of stability and convergence refer to the strong topology on Y and the weak topology on X. In the special case of X and Y being Hilbert spaces we obtained stability and convergence with respect to the strong topologies. Additionally, we studied the rate of the convergence, meaning that we quantified the speed of x_α^δ approaching $x^\dagger$ compared to the distance between the measured data y^δ and the exact data y. To do that, we did not metrize the weak topology of X, but instead we assumed X being a Hilbert space. In fact, it was also necessary to assume that Y is a Hilbert space. Giving convergence rates, was only possible under additional assumptions on $x^\dagger$ and on the operator F. In this section we will see how these additional conditions kept evolving and will introduce the objects that allowed the generalization of the results to the case of X and Y being Banach spaces.

2.1 The General Penalty term and a New Nonlinearity Condition

This step was made by M.Burger and S.Osher in *Convergence Rates of Convex Variational regularization* [5] published in the year 2004. Their achievement in this paper is the reformulation of all results we have attained, including the convergence

N. Uesseler, *Parameter Identification for a Stochastic Partial Differential Equation in the Nonstationary Case*, BestMasters,
https://doi.org/10.1007/978-3-658-50344-4_2

rates result, to the more general situation, in which we consider X being a Banach space and a generalized penalty term in the Tikhonov functional. Simultaneously, notice that the weak topology on X will not play a role anymore, instead we can consider *some* topology τ_X that is weaker than the strong norm-topology of the space. Let us first see the generalization of the penalty term.

Let X be a Banach space and $J : X \to \mathbb{R}$ be a convex functional that is not infinite everywhere and fulfills the following conditions:

1. J is τ_X−lower semi-continuous, i.e. for a sequence $x_n \to x$ converging in the τ_X topology as $n \to \infty$, it holds

$$J(x) \leq \liminf_{n\to\infty} J(x_n).$$

2. The sublevelsets $M_\rho := \{x \in X : J(x) \leq \rho\}$ of J are nonempty for $\rho \geq 0$ and τ_X−compact, i.e every sequence $\{x_n\}$ contained in M_ρ has a subsequence that converges in M_ρ with respect to the topology τ_X.

For the following analysis we need to assume that the continuous operator $F : X \to Y$ is also continuous with respect to the topology τ_X. But given this assumption we get some generalizations of the known results:

According to remark 1.1 we see inmidiately that the existence of a so-called *J-minimizing solution* is guaranteed. This is, the problem

$$\textit{minimize} \quad J(x) \quad \textit{subject to} \quad F(x) = y$$

has a solution. Naturally, we now define the (generalized) Tikhonov functional as

$$T_\alpha^\delta(x) := \| F(x) - y^\delta \|^2 + \alpha J(x)$$

and considering remarks 1.2, 1.3 and 1.4 we also know that the regularized solution, i.e the minimizer of the Tikhonov functional, exists and depends continuously (refering to the strong topology on Y and τ_X on X) on the data y^δ as well as that it converges to the J−minimizing solution as $\alpha(\delta) \to 0$ if $\alpha(\delta)$ is chosen correctly.

The next step is the generalization of the distance measure that was used to quantify the speed of the convergence in X. Indeed, if X is assumed to be a Banach space, using a distance that is related to the topology τ_X might be more natural than the norm of the space since the regularization procedure is not related to the norm but to the topology τ_X exclusively. The *Bregman distance* is for this reason useful. It is defined as the set

$$\mathcal{D}_J(u, v) := \{J(u) - J(v) - (p, u - v) : p \in \partial J(v)\},$$

where $(.,.)$ is the dual pairing and $\partial J(v)$ is the subdifferential of J at v, which is the set

$$\partial J(v) := \{p \in X^\star : J(v) + (p, u - v) \leq J(u) \text{ for all } u \in X\}$$

and an element $d \in \mathcal{D}_J(u, v)$ represents a distance between u and v after choosing an element $p \in \partial J(v)$, it has the form

$$\mathcal{D}_J^p(u, v) := d = J(u) - J(v) - (p, u - v).$$

Based on this concept of distance we now can focus on the corresponding result on convergence rates by M.Burger and S.Osher. The first task is to rewrite the source condition we discussed in the last section in terms of subdifferentiability. As we did in last section, we understand the $J-$minimizing solution $x^\dagger$ as the solution of a minimization problem and write the corresponding Lagrangian analogous to (1.5):

$$L(\omega, x) = J(x) + (\omega, F(x) - y).$$

Now the strong duality assumption is equivalent to the existence of a Lagrange multiplier ω, such that $(\omega, x^\dagger)$ is saddle point of the Lagrangian, which combining convexity of J and differentiability of F implies the (generalized) source condition, namely

$$- F'(x^\dagger)^\# \omega \in \partial J(x^\dagger).$$

Let us recall that in the convergence rate result that we already know (theorem 1.5) the source condition that we assumed on $x^\dagger$ was not sufficient, in fact, we also assumed Lipschitz continuity of F'. This condition will be replaced by a new nonlinearity condition on F. Let us put all the necessary ingridients together. $(.,.)$ corresponds to the dual pairing in X and the scalar product in Y and $\mathcal{B}_r$ is a ball of radius r.

Assumption 2.1 *X is a Banach space carrying also a weaker topology τ_X, Y a Hilbert space and $F : X \to Y$ is continuous w.r.t τ_X and Fréchet-differentiable with convex domain $\mathcal{D}(F)$. $x^\dagger$ is a $J-$minimizing solution that fulfills a source condition, i.e there exists $\omega \in Y$ satisfying*

$$F'(x^\dagger)^{\#}\omega \in \partial J(x^\dagger).$$

F fulfills the following nonlinearity condition: There exist $\eta > 0$ and $r > 0$ with

$$(F(z) - F(x^\dagger) - F'(x^\dagger)(z - x^\dagger), \omega) \le \eta\|F(x^\dagger) - F(z)\|\,\|\omega\| \quad \text{for all } z \in X.$$

Theorem 2.1 *Let assumption 2.1 hold and let $y^\delta \in Y$ be measurements of y with $\|y^\delta - y\| \le \delta$. Then for every minimizer x_α^δ of the Tikhonov functional T_α^δ there exists a $d \in \mathcal{D}_J(x_\alpha^\delta, x^\dagger)$ such that the choice $\alpha \sim \delta$ implies $d \sim \delta$.*

Proof Let $p := F'(x^\dagger)^{\#}\omega$ and $d := J(x_\alpha^\delta) - J(x^\dagger) - (p, x_\alpha^\delta - x^\dagger) \in \mathcal{D}_J(x_\alpha^\delta, x^\dagger)$. By definition of x_α^δ as a minimizer of T_α^δ it holds

$$\frac{1}{\alpha}\|F(x_\alpha^\delta) - y^\delta\|^2 + J(x_\alpha^\delta) \le \frac{1}{\alpha}\|F(x^\dagger) - y^\delta\|^2 + J(x^\dagger)$$

$$\le \frac{1}{\alpha}\delta^2 + J(x^\dagger).$$

Thus we get:

$$\frac{1}{\alpha}\|F(x_\alpha^\delta) - y^\delta + \frac{\alpha}{2}\omega\|^2 + d$$

$$= \frac{1}{\alpha}\|F(x_\alpha^\delta) - y^\delta\|^2 + (F(x_\alpha^\delta) - y^\delta, \omega) + \frac{\alpha}{4}\|\omega\|^2$$

$$\quad + J(x_\alpha^\delta) - J(x^\dagger) - (p, x_\alpha^\delta - x^\dagger)$$

$$= \frac{1}{\alpha}\|F(x_\alpha^\delta) - y^\delta\|^2 + J(x_\alpha^\delta) - J(x^\dagger) + \frac{\alpha}{4}\|\omega\|^2$$

$$\quad + (F(x_\alpha^\delta) - y^\delta, \omega) - (\omega, F'(x^\dagger)(x_\alpha^\delta - x^\dagger))$$

$$\le \frac{1}{\alpha}\delta^2 + J(x^\dagger) - J(x^\dagger) + \frac{\alpha}{4}\|\omega\|^2 + (F(x_\alpha^\delta) - y^\delta - F'(x^\dagger)(x_\alpha^\delta - x^\dagger), \omega)$$

$$= \frac{1}{\alpha}\delta^2 + \frac{\alpha}{4}\|\omega\|^2 + (F(x_\alpha^\delta) - y + y - y^\delta - F'(x^\dagger)(x_\alpha^\delta - x^\dagger), \omega)$$

$$= \frac{1}{\alpha}\delta^2 + \frac{\alpha}{4}\|\omega\|^2 + (F(x_\alpha^\delta) - F(x^\dagger) - F'(x^\dagger)(x_\alpha^\delta - x^\dagger), \omega) + (F(x^\dagger) - y^\delta, \omega)$$

$$\le \frac{1}{\alpha}\delta^2 + \frac{\alpha}{4}\|\omega\|^2 + \eta\|F(x_\alpha^\delta) - F(x^\dagger)\|\,\|\omega\| + \delta\|\omega\|$$

The claim would now follow by ignoring the first positive term in left-hand side if we could estimate $\|F(x_\alpha^\delta) - F(x^\dagger)\|$ and show that it behaves as δ for the choice $\alpha \sim \delta$. This is the last step. From the basic inequality $(a+b)^2 \le 2a^2 + 2b^2$ we see

$$
\begin{aligned}
\frac{1}{2}\|F(x_\alpha^\delta) - F(x^\dagger)\|^2 &\le \delta^2 + \|F(x_\alpha^\delta) - y^\delta\|^2 \\
&\le \delta^2 + \alpha J(x^\dagger) - \alpha J(x_\alpha^\delta) \\
&= \delta^2 - \alpha(F'(x^\dagger)^\# \omega, x_\alpha^\delta - x^\dagger) - \alpha d \\
&= \delta^2 - \alpha(F'(x^\dagger)(x_\alpha^\delta - x^\dagger) + F(x_\alpha^\delta) - F(x^\dagger), \omega) \\
&\quad - (F(x_\alpha^\delta) - F(x^\dagger), \omega) - \alpha d \\
&\le \delta^2 + \alpha\|\omega\|\eta\|F(x_\alpha^\delta) - F(x^\dagger)\| \\
&\quad + \alpha\|\omega\|\|F(x_\alpha^\delta) - F(x^\dagger)\| - \alpha d \\
&\le \delta^2 + \alpha\|\omega\|(\eta + 1)\|F(x_\alpha^\delta) - F(x^\dagger)\|,
\end{aligned}
$$

which implies

$$
\frac{1}{2}\|F(x_\alpha^\delta) - F(x^\dagger)\|^2 - \alpha\|\omega\|(\eta + 1)\|F(x_\alpha^\delta) - F(x^\dagger)\| - \delta^2 \le 0.
$$

By calculating the zeros of the quadratic function depending on $\|F(x_\alpha^\delta) - F(x^\dagger)\|$ in the left, we can see that

$$
\|F(x_\alpha^\delta) - F(x^\dagger)\| \le \alpha\|\omega\|(\eta + 1) + \sqrt{(\alpha\|\omega\|(\eta + 1))^2 + \delta^2}.
$$

Choosing $\alpha \sim \delta$ the claim follows. $\qquad\square$

2.2 Yet Another Nonlinearity Condition

In the year 2006 E.Resmerita and O.Scherzer [23] introduced yet another condition on the Operator F that replaces the nonlinearity conditions used in theorem 1.5 and in theorem 2.1. In principle, what these conditions do is controlling the error of the linear approximation of F at the ground truth $x^\dagger$. On the other hand, the source condition on $x^\dagger$ still has the same form that arrives from the assumption of strong duality in the dual problem. The Fréchet-differentiability could be relaxed by Gateâux-differentiability of F but we will stick to the first one to simplify the analysis. An important generalization takes place here in fact, namely Y may be now a Banach space and not necessarily a Hilbert space. $(.,.)$ refers to the dual pairing in the Banach space. Again, we first put the assumptions together.

Assumption 2.2 *X and Y are Banach spaces, both carrying also weaker topologies τ_X and τ_Y. $F : X \to Y$ is continuous w.r.t. τ_X and τ_Y and Fréchet-differentiable with convex domain $\mathcal{D}(F)$.*

$x^\dagger$ is a J−minimizing solution that fulfills a source condition, i.e. there exists $\omega \in Y^\star$ satisfying

$$F'(x^\dagger)^\# \omega \in \partial J(x^\dagger).$$

F fulfills the following nonlinearity condition: There exist $\eta > 0$ and $r > 0$ such that

$$\|F(x) - F(x^\dagger) - F'(x^\dagger)(x - x^\dagger)\| \leq \eta \mathcal{D}_J^\xi(x, x^\dagger) \qquad \text{for all } x \in X.$$

The Lagrange multiplier ω is small enough, namely

$$\eta \|\omega\| < 1.$$

Theorem 2.2 *Let assumption 2.2 hold and let $y^\delta \in Y$ be measurements of y with $\|y^\delta - y\| \leq \delta$. Let x_α^δ be a minimizer of the Tikhonov functional T_α^δ. Then for the choice $\alpha \sim \delta$ it holds*

$$\|F(x_\alpha^\delta) - F(x^\dagger)\| \sim \delta$$

and

$$\mathcal{D}_J^\xi(x_\alpha^\delta, x^\dagger) \sim \delta.$$

Before going into the proof, let us remark that the new nonlinearity condition on F arrives very naturally from the previous results. To see that, we first point out that the Bregman distance of the functional $J(x) = \|x\|^2$ defined on a Hilbert space is given by

$$\mathcal{D}_J^\xi(u, v) = J(v) - J(u) - (\xi, v - u)$$
$$= (v, v) - (u, u) - (\xi, v - u)$$
$$= (v, v) - (u, u) - (2u, v - u)$$
$$= (v, v) - 2(u, v) + (u, u)$$
$$= (v - u, v - u)$$
$$= \|v - u\|^2$$

since the only element in $\partial J(u)$ is $(2u, .)$. In particular, in this context the nonlinearity condition would take the form

$$\|r_\alpha^\delta\| := \|F(x_\alpha^\delta) - F(x^\dagger) - F'(x^\dagger)(x_\alpha^\delta - x^\dagger)\| \leq \eta \|x_\alpha^\delta - x^\dagger\|^2. \tag{2.1}$$

Now recall theorem 1.5. The nonlinearity condition that is assumed there is Lipschitz continuity of F'. But from the proof one can see that this condition is only necessary to prove exactly (2.1). In other words, the new condition is just a generalization of this necessary step in the mentioned proof.

Proof of theorem 2.2. By the minimizing property of x_α^δ we have

$$\frac{1}{2}\|F(x_\alpha^\delta) - y^\delta\|^2 \leq \frac{1}{2}\|F(x^\dagger) - y^\delta\|^2 + \alpha J(x^\dagger) - \alpha J(x_\alpha^\delta)$$
$$\leq \frac{1}{2}\delta^2 + \alpha J(x^\dagger) - \alpha J(x_\alpha^\delta).$$

Moreover, by the basic inequality $\frac{1}{2}(a + b)^2 \leq a^2 + b^2$ it holds

$$\frac{1}{2}\|F(x_\alpha^\delta) - y\|^2 \leq \frac{1}{2}(\|F(x_\alpha^\delta) - y^\delta\| + \|y^\delta - y\|)^2$$
$$\leq \|F(x_\alpha^\delta) - y^\delta\|^2 + \|y^\delta - y\|^2$$
$$= \|F(x_\alpha^\delta) - y^\delta\|^2 + \delta^2.$$

Combining these two inequalities with the definition of the Bregman distance we see

$$\frac{1}{4}\|F(x_\alpha^\delta) - F(x^\dagger)\|^2 \le \frac{1}{2}\delta^2 + \frac{1}{2}\|F(x_\alpha^\delta) - y^\delta\|^2$$

$$\le \delta^2 + \alpha J(x^\dagger) - \alpha J(x_\alpha^\delta)$$

$$= \delta^2 - \alpha(F'(x^\dagger)^\#\omega, x_\alpha^\delta - x^\dagger) - \alpha D_J^\xi(x_\alpha^\delta, x^\dagger)$$

$$\le \delta^2 + \alpha\|\omega\|\,\|F'(x^\dagger)(x_\alpha^\delta - x^\dagger)\| - \alpha D_J^\xi(x_\alpha^\delta, x^\dagger)$$

$$\le \delta^2 + \alpha\|\omega\|\,\|F(x_\alpha^\delta) - F(x^\dagger) - F'(x^\dagger)(x_\alpha^\delta - x^\dagger)\|$$

$$+ \alpha\|\omega\|\,\|F(x_\alpha^\delta) - F(x^\dagger)\| - \alpha D_J^\xi(x_\alpha^\delta, x^\dagger)$$

$$\le \delta^2 - \alpha\underbrace{(1 - \eta\|\omega\|)}_{>0}D_J^\xi(x_\alpha^\delta, x^\dagger) + \alpha\|\omega\|\,\|F(x_\alpha^\delta) - F(x^\dagger)\|,$$

where the last inequality uses the assumption. From this we get two inequalities if we first ignore the negative term in the right-hand side and then we ignore the left-hand side instead, namely

$$\frac{1}{4}\|F(x_\alpha^\delta) - F(x^\dagger)\|^2 \le \delta^2 + \alpha\|\omega\|\,\|F(x_\alpha^\delta) - F(x^\dagger)\| \tag{2.2}$$

and

$$\alpha(1 - \eta\|\omega\|)D_J^\xi(x_\alpha^\delta, x^\dagger) \le \delta^2 + \alpha\|\omega\|\,\|F(x_\alpha^\delta) - F(x^\dagger)\|. \tag{2.3}$$

(2.2) is equivalent to the inequality

$$\frac{1}{4}\|F(x_\alpha^\delta) - F(x^\dagger)\|^2 - \alpha\|\omega\|\,\|F(x_\alpha^\delta) - F(x^\dagger)\| - \delta^2 \le 0.$$

Understanding the left-hand side as a quadratic function in $\|F(x_\alpha^\delta) - F(x^\dagger)\|$ and solving for the zeros this implies directly

$$\|F(x_\alpha^\delta) - F(x^\dagger)\| \le 2\alpha\|\omega\| + 2(\alpha^2\|\omega\|^2 + \delta^2)^{\frac{1}{2}}.$$

Now with this estimate together with (2.3) we obtain

$$D_J^\xi(x_\alpha^\delta, x^\dagger) \le \frac{2}{1 - \eta\|\omega\|}[\delta^2/2\alpha + \alpha\|\omega\|^2 + \|\omega\|(\alpha^2\|\omega\|^2 + \delta^2)^{\frac{1}{2}}].$$

Finally choosing $\alpha \sim \delta$ we see that the claim holds. $\qquad\square$

2.3   The Appearance of the Variational Source Condition

The conditions we have studied, namely nonlinearity conditions on F and source conditions on the ground truth $x^\dagger$, assume implicitly differentiability of F at $x^\dagger$. In 2007 B.Hofmann, B.Kaltenbacher, C.Pöschl and O.Scherzer managed to reformulate these conditions as a variational inequality that does not require differentiability [14]. We will first see the new so called variational source condition and then study its relation with the already known conditions. Here the Tikhonov functional will be again slightly generalized:

$$T_\alpha^\delta(x) := \| F(x) - y^\delta \|_Y^p + \alpha J(x), \ \textit{for } p \geq 1 \tag{2.4}$$

Assumption 2.3 *X and Y are Banachspaces, both carrying also weaker topologies τ_X and τ_Y. $F : X \to Y$ is continuous w.r.t τ_X and τ_Y and has convex domain $\mathcal{D}(F)$.*
  $x^\dagger$ is a $J-$minimizing solution and there exist numbers $0 \leq \beta_1 < 1$, $0 \leq \beta_2$ and $\xi \in \partial J(x^\dagger)$ and some $r > 0$, such that

$$(\xi, x^\dagger - x) \leq \beta_1 \mathcal{D}_J^\xi(x, x^\dagger) + \beta_2 \| F(x) - F(x^\dagger) \| \qquad \textit{for all } x \in X.$$

Remark 2.1 *In the special case where assumption (2.1) and in particular differentiability of F is given, the following holds*

$$\begin{aligned}
(\xi, x - x^\dagger) &= (F'(x^\dagger)^\# \omega, x - x^\dagger) \\
&= (\omega, F'(x^\dagger)(x - x^\dagger)) \\
&= (\omega, F'(x^\dagger)(x - x^\dagger) + F(x^\dagger) - F(x) - F(x^\dagger) + F(x)) \\
&= (\omega, F'(x^\dagger)(x - x^\dagger) + F(x^\dagger) - F(x)) + (\omega, F(x) - F(x^\dagger)) \\
&\leq \eta \|\omega\| \| F(x^\dagger) - F(x) \| + \|\omega\| \| F(x) - F(x^\dagger) \| \\
&= \underbrace{(1 + \eta)\|\omega\|}_{=:\beta_2} \| F(x^\dagger) - F(x) \|.
\end{aligned}$$

and we recover assumption (2.3).

Similarly if assumption (2.2) is given, we obtain the following:

$$
\begin{aligned}
(\xi, x - x^\dagger) &= (F'(x^\dagger)^\# \omega, x - x^\dagger) \\
&= (\omega, F'(x^\dagger)(x - x^\dagger)) \\
&= (\omega, F'(x^\dagger)(x - x^\dagger) + F(x^\dagger) - F(x) - F(x^\dagger) + F(x)) \\
&= (\omega, F'(x^\dagger)(x - x^\dagger) + F(x^\dagger) - F(x)) + (\omega, F(x) - F(x^\dagger)) \\
&\leq \|\omega\| \, \|F'(x^\dagger)(x - x^\dagger) + F(x^\dagger) - F(x)\| + \|\omega\| \, \|F(x) - F(x^\dagger)\| \\
&\leq \underbrace{\eta \|\omega\| \, \mathcal{D}_J^\xi (x, x^\dagger)}_{=:\beta_1} + \underbrace{\|\omega\|}_{=:\beta_2} \|F(x) - F(x^\dagger)\|,
\end{aligned}
$$

which is assumption (2.3) again.

Finally recall that assumption (1.1) could already be replaced by assumption (2.2).

Altogether we see that assumption (2.3) covers all other conditions we had studied in their respective context. In other words the following theorem can be applied in all the former situations.

Theorem 2.3 *Let assumption 2.3 be fulfilled and let $y^\delta \in Y$ be measurements of y with $\|y^\delta - y\| \leq \delta$. Let x_α^δ be a minimizer of the Tikhonov functional T_α^δ in (2.4), where $p > 1$. Then for the choice $\alpha \sim \delta^{p-1}$ it holds*

$$\|F(x_\alpha^\delta) - F(x^\dagger)\| \sim \delta$$

and

$$\mathcal{D}_J^\xi (x_\alpha^\delta, x^\dagger) \sim \delta.$$

The version of the theorem in the original source adresses also the case $p = 1$, but we restrict our attention to $p > 1$ for simplicity.

Proof of theorem 2.3. By definition of x_α^δ it holds

$$\|F(x_\alpha^\delta) - y^\delta\|^p + \alpha J(x_\alpha^\delta) \leq \|F(x^\dagger) - y^\delta\|^p + \alpha J(x^\dagger)$$

and adding $\alpha \mathcal{D}_J^\xi (x_\alpha^\delta, x^\dagger)$ on both sides, using the definition of the Bregman distance and then the variational source condition yields

$$\|F(x_\alpha^\delta) - y^\delta\|^p + \alpha \mathcal{D}_J^\xi(x_\alpha^\delta, x^\dagger)$$

$$\leq \delta^p + \alpha(J(x^\dagger) - J(x_\alpha^\delta) + \mathcal{D}_J^\xi(x_\alpha^\delta, x^\dagger))$$

$$= \delta^p - \alpha(\xi, x_\alpha^\delta - x^\dagger)$$

$$\leq \delta^p + \alpha(\beta_1 \mathcal{D}_J^\xi(x_\alpha^\delta, x^\dagger) + \beta_2 \|F(x_\alpha^\delta) - F(x^\dagger)\|)$$

$$\leq \delta^p + \alpha(\beta_1 \mathcal{D}_J^\xi(x_\alpha^\delta, x^\dagger) + \beta_2(\|F(x_\alpha^\delta) - y^\delta\| + \delta)) \qquad (2.5)$$

$$= \delta^p + \alpha\beta_1 \mathcal{D}_J^\xi(x_\alpha^\delta, x^\dagger) + \alpha\beta_2\|F(x_\alpha^\delta) - y^\delta\| + \alpha\beta_2\delta.$$

Rearranging the terms we get

$$\|F(x_\alpha^\delta) - y^\delta\|^p - \alpha\beta_2\|F(x_\alpha^\delta) - y^\delta\| + \alpha(1 - \beta_1)\mathcal{D}_J^\xi(x_\alpha^\delta, x^\dagger)$$
$$\leq \delta^p + \alpha\delta\beta_2. \qquad (2.6)$$

First we ignore the third term in the left hand side of (2.6) to see

$$\|F(x_\alpha^\delta) - y^\delta\|^p - \underbrace{\alpha\beta_2}_{b} \underbrace{\|F(x_\alpha^\delta) - y^\delta\|}_{a} \leq \delta^p + \alpha\delta\beta_2.$$

Using the Young inequality

$$ab \leq \frac{a^p}{p} + \frac{b^q}{q}, \quad \text{where } \frac{1}{p} + \frac{1}{q} = 1$$

we obtain

$$\alpha\beta_2\|F(x_\alpha^\delta) - y^\delta\| \leq \frac{1}{q}(\alpha\beta_2)^q + \frac{1}{p}\|F(x_\alpha^\delta) - y^\delta\|^p$$

and thus

$$\|F(x_\alpha^\delta) - y^\delta\|^p \leq \delta^p + \alpha\delta\beta_2 + \frac{1}{q}(\alpha\beta_2)^q + \frac{1}{p}\|F(x_\alpha^\delta) - y^\delta\|^p$$

$$\iff \frac{1}{q}\|F(x_\alpha^\delta) - y^\delta\|^p \leq \delta^p + \alpha\delta\beta_2 + \frac{1}{q}(\alpha\beta_2)^q$$

$$\iff \|F(x_\alpha^\delta) - y^\delta\|^p \simeq \delta^p$$

$$\iff \|F(x_\alpha^\delta) - y^\delta\| \simeq \delta,$$

where we used the choice $\alpha \sim \delta^{p-1}$. Now we go back to (2.6), but this time we ignore the first term on the left hand side. Here we use the Young inequality in the same way as before as well as the choice $\alpha \sim \delta^{p-1}$

$$\alpha(1 - \beta_1)\mathcal{D}_J^\xi(x_\alpha^\delta, x^\dagger) \leq \delta^p + \alpha\delta\beta_2 + \alpha\beta_2 \| F(x_\alpha^\delta) - y^\delta \|$$

$$\leq \delta^p + \alpha\delta\beta_2 + \frac{1}{q}(\alpha\beta_2)^q + \frac{1}{p}\| F(x_\alpha^\delta) - y^\delta \|^p$$

$$\simeq \delta^p,$$

which implies again using $\alpha \sim \delta^{p-1}$

$$\mathcal{D}_J^\xi(x_\alpha^\delta, x^\dagger) \simeq \delta.$$

$\square$

2.4　Generalizing Variational Source Conditions

B. Hofmann himself in joint work with M. Yamamoto point out a slight generalization of theorem 2.3 in the paper *'On the interplay of source conditions and variational inequalities for nonlinear ill-posed problems'* [15] published in 2009. Namely the variational inequality is relaxed. This new assumption reads as follows.

Assumption 2.4 *X and Y are Banach spaces, both carrying also weaker topologies τ_X and τ_Y. $F : X \to Y$ is continuous w.r.t τ_X and τ_Y and has convex domain $\mathcal{D}(F)$.*

$x^\dagger$ is a J−minimizing solution and there exist numbers $0 \leq \beta_1 < 1, 0 \leq \beta_2$ and $0 < \kappa \leq 1$, as well as $\xi \in \partial J(x^\dagger)$ and some $r > 0$, such that

$$\langle \xi, x^\dagger - x \rangle \leq \beta_1 \mathcal{D}_J^\xi(x, x^\dagger) + \beta_2 \| F(x) - F(x^\dagger) \|^\kappa \qquad \textit{for all } x \in X.$$

This assumption only differs from assumption 2.3 in the exponent κ in the right hand side of the inequality. Now the corresponding result, whose proof is analogous to the one of theorem 2.3 and hence also based on the Young inequality.

Theorem 2.4 *Let assumption 2.4 be fulfilled and let $y^\delta \in Y$ be mesurements of y with $\| y^\delta - y \| \leq \delta$. Let x_α^δ be a minimizer of the Tikhonov functional T_α^δ in (2.4), where $p > 1$. Then for the choice $\alpha \sim \delta^{p-\kappa}$ it holds*

$$\mathcal{D}_J^\xi(x_\alpha^\delta, x^\dagger) \sim \delta^\kappa.$$

This result suggests that the variational inequality as well as the Tikhonov functional, which are based on the norm $\|.\|_Y$ of the Banach space Y could be relaxed and still provide rates of convergence. In fact, the same year B.Hofmann together with R.Bot considered the following assumption [2].

Assumption 2.5   *X and Y are Banach spaces, both carrying also weaker topologies τ_X and τ_Y. $F : X \to Y$ is Gâteaux-differentiable and continuous w.r.t τ_X and τ_Y and has convex domain $\mathcal{D}(F)$.*

ψ and ϕ are index functions, i.e strictly increasing functions on $[0, \infty)$ with $\psi(0) = 0 = \phi(0)$. Both are twice differentiable in the interior of their domain, ψ is strictly convex and ϕ is concave. The Tikhonov functional is defined by

$$T_\alpha^\delta(x) := \psi(\|F(x) - y^\delta\|) + \alpha J(x), \text{ for } p \geq 1,$$

where J is Gâteaux-differentiable. $x^\dagger$ is a $J-$minimizing solution and there exist numbers $0 \leq \beta_1 < 1, 0 \leq \beta_2$ and $\xi \in \partial J(x^\dagger)$ and some $r > 0$, such that

$$(\xi, x^\dagger - x) \leq \beta_1 \mathcal{D}_J^\xi(x, x^\dagger) + \beta_2 \phi(\|F(x) - F(x^\dagger)\|) \qquad \text{for all } x \in X.$$

Theorem 2.5   *Let assumption 2.5 be fulfilled and let $y^\delta \in Y$ be measurements of y with $\|y^\delta - y\| \leq \delta$. Let x_α^δ be a minimizer of the Tikhonov functional in the assumption. Then for*

$$\alpha \sim \frac{1}{\beta_2} \frac{\psi'}{\phi'}(\delta)$$

one obtains the rate of convergence

$$\mathcal{D}_J^\xi(x_\alpha^\delta, x^\dagger) \sim \phi(\delta).$$

Instead of proving this result, we will first generalize the Tikhonov functional allowing different types of fidelity terms and then we will reformulate a convergence result for that situation, which will prove to be more useful for us.

Remark 2.2 (Local nonlinearity conditions). *For simplicity we assumed for the results above the nonlinearity conditions to hold in the whole space X. This is in the literature not required. It is sufficient that the nonlinearity condition holds only on a sublevelset of $T_{\bar{\alpha}}^{\delta}$ for some fixed but arbitrary $\bar{\alpha}$ and correspondingly chosen level [14, Rem. 3.6].*

To this point we have been working with a version of the Tikhonov functional in which the fidelity term is based on the norm of the Banach space Y. Here we want to generalize this object and allow for similarity measures that are useful for a bigger class of ill-posed problems. This generalized fidelity term should still have the properties that ensure the results we already know. The triangle inequality of the norm turns out to be a delicate topic that needs special attention when studying rates of convergence. We will present the important example of the Kullback–Leibler divergence as a similarity measure.

3.1 The Essential Properties

We are interested in different ways to determine the similarity between measured data and reconstruction. In the Tikhonov functional, the term that has this responsibility is the fidelity term. To this point we have used $\| F(x) - y \|^2$, $\| F(x) - y \|^p$ for $p > 1$ or $\psi(\| F(x) - y \|)$ for some strictly convex function ψ. Our task now is to generalize this object in a meaningful way. C.Pöschl in her dissertation [22] in 2008 systematically reformulates the basic results we have seen, minimizing the assumptions on the penalty term, the fidelity term, the topologies chosen on the spaces X and Y and the operator $F : X \to Y$. In remarks 1.1, 1.2, 1.3 and 1.4 we have already highlighted the properties that were used in the proofs regarding

1. Existence of a J−minimizing solution x_0
2. Existence of the minimizer of T_α^δ
3. Stability of the regularized solution x_α^δ
4. Convergence of x_α^δ to x_0

© The Author(s), under exclusive license to Springer Fachmedien Wiesbaden GmbH, part of Springer Nature 2025

N. Uesseler, *Parameter Identification for a Stochastic Partial Differential Equation in the Nonstationary Case*, BestMasters,
https://doi.org/10.1007/978-3-658-50344-4_3

We should also remark that we managed to prove these results **without** using the triangle inequality of the norm. However we will present a slightly more general setting than the one in the dissertation of C.Pöschl. Namely the one presented by J.Flemming, who, inspired in fact by C.Pöschl, reformulated the assumptions in such a way that the measured data and the exact data do not have to live in the same space [10]. This will prove to be more useful for us in the future. The proofs can be found in the original sources, but note that they use the same techniques that are used in the proofs we saw. In principle, the following sufficient conditions were extracted carefully of the known proofs.

Assumption 3.1 *Consider the forward operator $F : X \to Y$, where X and Y are Banach spaces carrying topologies τ_X and τ_Y, such that F is sequentially continuous with respect to τ_X and τ_Y. We also assume that $\mathcal{D}(F)$ is closed in τ_X. Moreover, we assume attainability, meaning, for the exact data y, there exists a solution $x^\dagger$ of $F(x) = y$. The measured data lives in the Banach space Z, which also carries a topology τ_Z.*

Let us define the (generalized) Tikhonov functional as

$$T_\alpha^z(x) := S(F(x), z) + \alpha J(x).$$

Here, $J : X \to [0, \infty]$ is assumed to be proper and lower semicontinuous with respect to τ_X. For $S : Y \times Z \to \mathbb{R}$ the following assumptions hold true:

- *S is sequentially lower semicontinuous with respect to $\tau_Y \otimes \tau_Z$.*
- *$S(y, z_n) \to 0$ implies there exists a $z \in Z$, such that $z_n \to z$ in τ_Z.*
- *$z_n \to z$ in τ_Z implies $S(y, z_n) \to S(y, z)$ for all $y \in Y$ with $S(y, z) < \infty$ and hence $S(y, z_n)$ is a bounded sequence.*

Finally, the condition that replaces the use of Banach–Alouglu: For every $\alpha > 0$, $z \in Z$ and $M > 0$ the sublevel sets

$$\mathcal{M}_{\alpha,z}(M) := \{x \in \mathcal{D}(J) \cap \mathcal{D}(F) : T_\alpha^z(x) \leq M\}$$

are sequentially compact with respect to τ_X, that is, every sequence in $\mathcal{M}_{\alpha,y}(M)$ has a subsequence that converges with respect to τ_X.

Nevertheless, here we want to note that to achieve the different results we studied about

5. Rates of convergence

it was necessary to impose the triangle inequality on S. More specifically the result
that was reformulated for this setting, was theorem 2.3 together with its assumption
2.3. In the proof we can see that, in fact, the triangle inequality of the norm is used
(see inequality (2.5)). Under the following assumptions the theorem could be also
extended to the more general case.

Assumption 3.2 *Assumption 3.1 holds with $Y = Z$. J is also convex and S satisfies
a triangle inequality, i.e*

$$S(x_1, x_3) \leq S(x_1, x_2) + S(x_2, x_3).$$

*x_0 is a $J-$minimizing solution and there exist numbers $0 \leq \beta_1 < 1, 0 \leq \beta_2$
and $\xi \in \partial J(x_0)$, such that*

$$(\xi, x - x_0) \leq \beta_1 \mathcal{D}_J^\xi(x, x_0) + \beta_2 S(F(x), F(x_0)) \qquad \textit{for all } x \in X.$$

Remark 3.1 *Notice that the fidelity functional does not have to fulfill itself the
triangle inequality. If $S = \tilde{S}^p$ for some $p \geq 1$ is the fidelity functional, it is sufficient
that $\tilde{S}$ fulfills the triangle inequality. This is the case for example when $S(y', y) =
\|y' - y\|^2$.*

Instead of formulating the corresponding result and showing its proof (which is
analogous to theorem 2.3 and its proof), we will modify the noise model we are
using (it is in the noise measuring, where the triangle inequality plays a role) and
will reformulate in that setting the theorem. We will see the proof and will see that,
in fact, it covers the latter situation.

3.2   An Alternative Noise Model

Up to this point, the model we have been using to describe the noise in our measur-
ments y^δ was

$$S(y, y^\delta) \leq \delta, \tag{3.1}$$

where y is the exact data and $\delta > 0$ represented a known bound of the error. This model had the advantage, that in combination with the triangle inequality of S one could say for some element $z = F(x)$

$$S(z, y) \leq S(z, y^\delta) + S(y, y^\delta)$$
$$\leq S(z, y^\delta) + \delta.$$

Then it was possible to estimate the first term on the right-hand side. This is how the triangle inequality played a role in all the theorems on convergence rates we have seen.

However, if the triangle inequality does not hold, the bound $S(y, y^\delta) \leq \delta$ is not sufficient. To handle this case F. Werner [26] defined in his dissertation in 2011, motivated by an inverse problem with Poisson data, an alternative noise model, which can be used when the triangle inequality is not available. In this setting it might be necessary to allow a different functional, say $\mathcal{T}(z, y)$, to measure the similarity between the exact data y and the reconstructions $z = F(x)$. We also assume $\mathcal{T} \geq 0$ and $\mathcal{T}(y, y) = 0$. It should be possible to estimate $\mathcal{T}(z, y)$ based on information about $S(z, y^\delta)$ and then it should be possible as well to estimate the similarity of x_α^δ with the ground truth x_0 based on information about $\mathcal{T}(z, y)$. The functional S will still be the functional that measures the reconstructions z against the observation y^δ and the one that is minimized in the Tikhonov functional to define the regularization x_α^δ. But measuring the similarity between z and the exact data y, which is necessary to estimate the convergence speed, will not be the function of S anymore.

Let us define

$$\mathrm{Err}(z)(y, y^\delta) := \mathcal{T}(z, y) - S(z, y^\delta) + S(y, y^\delta) \tag{3.2}$$

and assume that we know that this term can be bounded by a constant that does not depend on z, namely

$$\mathrm{Err}(z)(y, y^\delta) \leq \delta. \tag{3.3}$$

The idea here is still being able to bound the term $\mathcal{T}(z, y)$ based on data $S(z, y^\delta)$. One possible interpretation of this definition of error can be seen by rewriting (3.3) as

$$\mathcal{T}(z, y) - \delta \leq S(z, y^\delta) - S(y, y^\delta).$$

The right hand side is S corrected in such a way that it vanishes when the reconstruction z coincides with the exact data y and it gives us an upper bound for the similarity $T(z, y)$ up to an error δ. T should be as strong (or big) as possible to give good rates of convergence but needs to be weaker (or smaller) than the right hand side of the inequality in order to be estimated based on it. This condition might or might not be possible with $T = S$, which is the reason why we introduced T. Moreover, the fact that $T \geq 0$ enforces the term $S(z, y^\delta) - S(y, y^\delta)$ to stay above $-\delta$, thus the difference between the minimal value of $S(\cdot, y^\delta)$ and $S(y, y^\delta)$ is smaller than δ. In other words, y is close to being the optimal value of $S(\cdot, y^\delta)$, which can be understood as similarity between y and y^δ (Fig. 3.1).

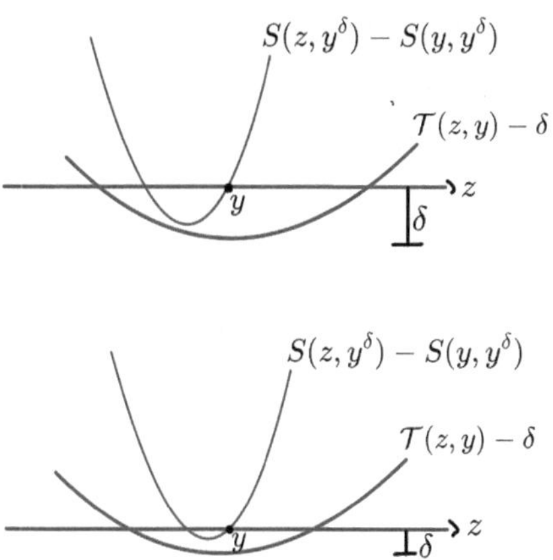

Fig. 3.1 The picture shows that for smaller δ, the minimal value of $S(\cdot, y^\delta)$ approaches $S(y, y^\delta)$. Thus for small δ, y is *almost* the best value of $S(\cdot, y^\delta)$. Depending on the specific properties of S, this can in turn imply that y and y^δ need to be similar

Example 3.1   Consider the Hilbert space Y and the fidelity term given by the square of the norm

$$S(z, w) := \|z - w\|^2,$$

which does not fulfill the triangle inequality. Set $T(z, w) = \frac{1}{2} S(z, w) = \frac{1}{2} \|z - w\|^2$
We have by the inequality $\frac{1}{2} a^2 - b^2 \le (a - b)^2$

$$\begin{aligned}
\mathrm{Err}(z)(y, y^\delta) &= T(z, y) - S(z, y^\delta) + S(y, y^\delta) \\
&= \frac{1}{2} \|z - y\|^2 - \|z - y^\delta\|^2 - \|y + y^\delta\|^2 \\
&\le (|\|z - y\| - \|z - y^\delta\||)^2 + \|y - y^\delta\|^2 \\
&\le \|y - y^\delta\|^2 + \|y - y^\delta\|^2 \\
&= 2\|y - y^\delta\|^2
\end{aligned}$$

*which can be bounded by δ if we have information given as in the old noise model
(3.1). In particular, the old noise model with quadratic fidelity term is generalized
by the new approach.*

Since the functional in the last example is norm based, we could have used the
triangle inequality from the beginning as in remark 3.1, so this example was not the
original motivation. The following example is, in fact, relevant for us.

*Example 3.2 Consider the space of probability measures $\mathcal{P}_1(\Omega)$ on some space
$\Omega \subset \mathbb{R}^d$ and the subset $\mathcal{Z}(\Omega) \subset \mathcal{P}_1(\Omega)$ of probability measures with a density with
respect to the Lebesgue measure $\mathrm{d}x$. The elements in $\mathcal{Z}(\Omega)$ will be directly identified
with its density. Consider now the nonsymmetric fidelity functional*

$$S : \mathcal{Z}(\Omega) \times \mathcal{P}_1(\Omega) \to \mathbb{R}$$

$$(z, G) \mapsto \begin{cases} -\int \ln z \, \mathrm{d}G - c & z > 0 \ \ a.e. \\ \infty & else, \end{cases}$$

*where $c := -\int \ln g^\dagger \, \mathrm{d}g^\dagger$ and $g^\dagger \in \mathcal{Z}(\Omega)$ is the exact data. Then by definition
it holds $S(g^\dagger, g^\dagger) = 0$. Let $g^m \in \mathcal{P}_1(\Omega)$ be a measurement. Now consider the
Kullback–Leibler divergence defined on $\mathcal{Z}(\Omega) \times \mathcal{Z}(\Omega)$*

$$\mathrm{KL}(g, \hat{g}) = \int (\hat{g} \ln \frac{\hat{g}}{g} + g - \hat{g}) \, \mathrm{d}x.$$

*Notice that if the second argument is the exact data $g^\dagger$, both functionals coincide,
namely*

$$
\begin{aligned}
\mathrm{KL}(g, g^{\dagger}) &= \int \left(g^{\dagger} \ln \frac{g^{\dagger}}{g} + g - g^{\dagger} \right) \mathrm{d}x \\
&= \int (g^{\dagger} \ln g^{\dagger})\, \mathrm{d}x - \int (g^{\dagger} \ln g)\, \mathrm{d}x + \underbrace{\int g\, \mathrm{d}x}_{=1} - \underbrace{\int g^{\dagger}\, \mathrm{d}x}_{=1} \\
&= - \int \ln g\, \mathrm{d}g^{\dagger} - c \\
&= S(g, g^{\dagger}),
\end{aligned}
$$

which motivates using KL *as the measure of similarity for exact data. Recall that* KL ≥ 0 *and* $\mathrm{KL}(g, g) = 0$ *for all* $g \in \mathcal{Z}(\Omega)$. *Now consider the error functional*

$$
\begin{aligned}
\mathrm{Err}(z)(g^{\dagger}, g^{m}) &= \mathrm{KL}(z, g^{\dagger}) - S(z, g^{m}) + S(g^{\dagger}, g^{m}) \\
&= - \int \ln z\, \mathrm{d}g^{\dagger} - c + \int \ln z\, \mathrm{d}g^{m} + c - \int \ln g^{\dagger}\, \mathrm{d}g^{m} - c \\
&= - \int \ln z\, \mathrm{d}g^{\dagger} + \int \ln \frac{z}{g^{\dagger}}\, \mathrm{d}g^{m} - c \\
&= - \int \ln z\, \mathrm{d}g^{\dagger} + \int \ln \frac{z}{g^{\dagger}}\, \mathrm{d}g^{m} + \int \ln g^{\dagger}\, \mathrm{d}g^{\dagger} \\
&= \int \ln \frac{z}{g^{\dagger}}\, (\mathrm{d}g^{m} - \mathrm{d}g^{\dagger}).
\end{aligned}
$$

Later we will see that this term can be bounded in fact by a constant (with sufficiently high probability). In our application S will show up as the negative log-likelihood functional with the modification of the constant c. As a last remark, we note that c in the fidelity term, and thus in the Tikhonov functional, is unknown, but this is for the minimization problem not relevant, since c is constant.

3.3 The General Convergence Rate Result

We are now ready to formulate and prove the most general result on rates of convergence we will study. We will assume the more general noise model we introduced and a very general version of the variational source inequality. Let us first see the assumption that will replace assumption 3.2. This result was published in *Convergence rates in expectation for Tikhonov-type regularization of inverse problems with Poisson data* by F.Werner and T.Hohage [27] in 2012.

Assumption 3.3 *Assumption 3.1 holds and J is also convex. There exists a constant $\delta \geq 0$ such*

$$\mathrm{Err}(z)(y, y^\delta) \leq \delta$$

independently of z, with Err *as defined in (3.2). $x^\dagger$ is a $J-$minimizing solution and there exists an index function ϕ (i.e. $\phi(0) = 0$, continuous and monotonically increasing) such that ϕ is concave, as well as a parameter $\beta > 0$ and $\xi \in \partial J(x^\dagger)$ and some $r > 0$, such that for all*

$$\beta \mathcal{D}_J^\xi(x, x^\dagger) \leq J(x) - J(x^\dagger) + \phi\left(\mathcal{T}(F(x), F(x^\dagger))\right) \quad \text{for all } x \in X. \quad (3.4)$$

Notice that the variational source condition is the same that we find in assumption 2.5. One parameter was absorbed by the function ϕ, and the definition of $\mathcal{D}_J^\xi(x, x^\dagger)$ was used.

In order to formulate and prove the result we are aiming for, we need the Fenchel conjugate of a function $\gamma : \mathbb{R} \to (-\infty, \infty]$, namely

$$\gamma^*(s) = \sup_{\tau \in \mathbb{R}}(s\tau - \gamma(\tau)).$$

We also need the facts that for convex and continuous γ Young's inequality

$$s\tau \leq \gamma(\tau) + \gamma^*(s) \text{ for all } s, \tau \in \mathbb{R}$$
$$s\tau = \gamma(\tau) + \gamma^*(s) \iff \tau \in \partial\gamma(s).$$

holds and

$$\phi^{**} = \phi.$$

Theorem 3.1 *Let assumption 3.3 be fulfilled and let x_α^δ be a minimizer of the Tikhonov functional in the assumption. Then for all $\alpha > 0$ we have*

$$\beta \mathcal{D}_J^\xi(x_\alpha^\delta, x^\dagger) \leq \frac{\delta}{\alpha} + (-\phi)^*\left(-\frac{1}{\alpha}\right).$$

Moreover, the optimal choice $\hat{\alpha}$, i.e. the minimizer of the right-hand side, is given if and only if

$$-\frac{1}{\hat{\alpha}} \in \partial(-\phi)(\delta),$$

which gives the optimal rate

$$\beta D_J^\xi(x_{\hat{\alpha}}^\delta, x^\dagger) \leq \phi(\delta).$$

Proof By definition of x_α^δ it holds

$$S(F(x_\alpha^\delta), y^\delta) + \alpha J(x_\alpha^\delta) \leq S(F(x^\dagger), y^\delta) + \alpha J(x^\dagger)$$
$$= S(y, y^\delta) + \alpha J(x^\dagger)$$

and thus

$$J(x_\alpha^\delta) - J(x^\dagger) \leq \frac{1}{\alpha}(S(y, y^\delta) - S(F(x_\alpha^\delta), y^\delta)).$$

Combining this with the variational inequality, the definition of δ and the definition of the fenchel conjugate we obtain

$$\beta D_J^\xi(x_\alpha^\delta, x^\dagger) \leq J(x_\alpha^\delta) - J(x^\dagger) + \phi\left(T(F(x_\alpha^\delta), y)\right)$$
$$= J(x_\alpha^\delta) - J(x^\dagger) + \phi\left(T(F(x_\alpha^\delta), y)\right)$$
$$\leq \frac{1}{\alpha}[S(y, y^\delta) - S(F(x_\alpha^\delta), y^\delta)] + \phi\left(T(F(x_\alpha^\delta), y)\right)$$
$$\leq \frac{1}{\alpha}[\delta - T(F(x_\alpha^\delta), y)] + \phi\left(T(F(x_\alpha^\delta), y), y)\right)$$
$$\leq \frac{\delta}{\alpha} - \frac{1}{\alpha}T(F(x_\alpha^\delta), y) + \phi\left(T(F(x_\alpha^\delta), y)\right)$$
$$\leq \frac{\delta}{\alpha} + \sup_{s \geq 0}(\frac{s}{-\alpha} - (-\phi(s)))$$
$$= \frac{\delta}{\alpha} + (-\phi)^*(-\frac{1}{\alpha}).$$

Now consider

$$\inf_{\alpha>0}\left[\frac{\delta}{\alpha}+(-\phi)^*(-\frac{1}{\alpha})\right] = -\sup_{\alpha>0}\left[-\frac{\delta}{\alpha}-(-\phi)^*(-\frac{1}{\alpha})\right]$$

$$= -\sup_{s<0}[\delta s - (-\phi)^*(s)]$$

$$= -(-\phi)^{**}(\delta)$$

$$= \phi(\delta).$$

Finally, the infimum is attained if and only if

$$\frac{\delta}{\alpha}+(-\phi)^*(-\frac{1}{\alpha}) = -(-\phi)^{**}(\delta)$$

$$\Longleftrightarrow \quad \delta(-\frac{1}{\alpha}) = (-\phi)^*(-\frac{1}{\alpha})+(-\phi)^{**}(\delta)$$

$$\Longleftrightarrow \quad -\frac{1}{\alpha} \in \partial(-\phi)(\delta),$$

by Young's inequality in the equality case. $\qquad\qquad\qquad\qquad\qquad\qquad\square$

The Tools to Work with Random Data

Now that we have a result on convergence rates based on the more general noise model, which does not rely on the triangle inequality of the fidelity functional, we can use the negative log-likelihood functional as fidelity term in the Tikhonov functional. The goal here is to estimate the solution of ill-posed problems, whose exact data is a probability density $g^\dagger$ and the given measured data consists of realizations of random variables with distribution $g^\dagger dx$. We will introduce a general concentration inequality from probability theory and combine it with the convergence rates result to obtain the desired method.

4.1 The General Random Setting

We consider the following problem. Let $\Omega \subset \mathbb{R}^d$ be a bounded Lipschitz domain and $v \in \mathcal{P}_1(\Omega)$ be a probability measure on it, which is absolutely continuous with respect to the Lebesgue measure dx, namely $v = g\,dx$. Let $\{Q_i\}_{i=1,\ldots,n}$ be a family of independent random variables, which are all $v-$distributed. Given the observations $Q_i = q_i$ we would like to estimate the underlying density $g^\dagger$. To do that consider the likelihood of g given the measurements q_i

$$\prod_i g(q_i)$$

and the log-likelihood

$$\sum_i \ln g(q_i).$$

N. Uesseler, *Parameter Identification for a Stochastic Partial Differential Equation in the Nonstationary Case*, BestMasters,
https://doi.org/10.1007/978-3-658-50344-4_4

If we define the empirical probability measure as

$$g^m := \frac{1}{n} \sum_i \delta_{q_i}, \tag{4.1}$$

(where m refers to *measurement*) we see that

$$-\frac{1}{n} \sum_i \ln g(q_i) = -\int \ln g \, dg^m.$$

This coincides up to a constant with the functional S in 3.2 and motivates its definition. Now consider a nonlinear operator

$$F : \mathcal{B} \to \mathcal{Z}(\Omega) \cap H^s(\Omega),$$

where $\mathcal{B}$ is a convex subset of a Banach space X, $\mathcal{Z}(\Omega) \subset \mathcal{P}_1(\Omega)$ is the subset of Lebesgue-absolutely continuous measures and $H^s(\Omega)$ is a Sobolev space with $s > \frac{d}{2}$. Notice that $H^s(\Omega)$ then embedds continuously in $C^0(\bar{\Omega})$ and thus in $L^\infty(\Omega)$ (see e.g. [1, Thm. 8.13]) We are interested in the respective nonlinear ill-posed problem

$$F(x) = g^\dagger,$$

where $g^\dagger \in \mathcal{Z}(\Omega) \cap H^s(\Omega)$ is the exact data and the measured data g^m as in (4.1) represents realizations of the density $g^\dagger$. In order to estimate x based on the empirical measure g^m, we will minimize the Tikhonov functional

$$T^n_\alpha(x) = S(F(x), g^m) + \alpha J(x) \tag{4.2}$$

over $x \in \mathcal{B}$, where S is the functional defined in example 3.2 and the superscript n denotes the number of realizations (we will see later that this number determines the error level) and J is convex. In the same example we also discussed the noise model that we will use. Finally, we will also assume that the range of F is bounded in $H^s(\Omega)$, namely

$$\sup_{x \in \mathcal{B}} \| F(x) \|_{H^s} \leq R < \infty. \tag{4.3}$$

Following the ideas of F.Dunker and T.Hohage, in *On Parameter Identification in Stochastic Differential Equations by Penalized Maximum Likelihood (2014)* [7], we will be able to show that this approach delivers minimizers, whose rate of convergence to the ground truth $x^\dagger$ can be estimated. But first, we have to introduce the main tool of probability theory that we need.

4.2   The Concentration Inequality

The Inequality that will be useful to us arrives from a result by P.Massart in 1999, which in our notation can be stated as follows [19]:

Theorem 4.1 *Let $\mathcal{F} \subset L^\infty(\Omega)$ be a countable familiy of functions with $\| f \|_\infty \leq b < \infty$ for all $f \in \mathcal{F}$. Let*

$$
Z := n \sup_{f \in \mathcal{F}} \left| \int f(\mathrm{d}g^m - g^\dagger \mathrm{d}x) \right|,
$$

where g^m is the empirical measure representing n realizations, and let

$$
V := n \sup_{f \in \mathcal{F}} \int f^2 g^\dagger \, \mathrm{d}x.
$$

Then

$$
\mathbb{P}\left[Z \geq (1 + \epsilon)\mathbb{E}[Z] + \sqrt{8V\gamma} + \kappa(\epsilon)b\gamma \right] \leq \exp\left(-\gamma\right)
$$

for all $\epsilon, \gamma > 0$ where $\kappa(\epsilon) = 2.5 + 32/\epsilon$.

This theorem helps us bound the deviation of the random variable Z from its mean with high probability. F.Dunker and T.Hohage managed to estimate $\mathbb{E}[Z]$ to prove the following corollary.

Corollary 4.1 *Let $B_s(R)$ be the Ball of radius R in $H^s(\Omega)$. There exists a constant $C_c \geq 1$ depending only on Ω and s such that for $\rho \geq RC_c$ and for all $n \in \mathbb{N}$ it holds*

$$
\mathbb{P}\left[\sup_{z \in B_s(R)} \left| \int z(\mathrm{d}g^m - g^\dagger \mathrm{d}x) \right| \geq \frac{\rho}{\sqrt{n}} \right] \leq \exp\left(-\frac{\rho}{RC_c}\right).
$$

At this point, let us recall the error functional derived for the fidelity term S in the Tikhonov functional (4.2). In example 3.2 we found out that

$$\left| \mathrm{Err}(z)(g^\dagger, g^m) \right| = \left| \int \ln \frac{z}{g^\dagger} \, (\mathrm{d}g^m - \mathrm{d}g^\dagger) \right|. \tag{4.4}$$

And from theorem 3.1 we know that (given a variational source condition) we can get convergence rates of the Tikhonov minimizer under the assumption that $\mathrm{Err}(z)(g^\dagger, g^m)$ is bounded independently of z by a certain error level. This bound has to be probabilistic, since g^m is a random variable and thus the convergence from theorem 3.1 will also be a probabilistic result. The utility of corollary 4.1 becomes clear now: establishing a probabilistic error bound for the measurement g^m given n realizations.

However, corollary 4.1 requires the functions under the integral to be bounded, but to get a bound on (4.4), we need to consider the logarithm of the quotient of two densities in $B_s(R)$, which could diverge to $-\infty$ or not be well defined. In fact, it is actually not possible to bound (4.4) using directly corollary 4.1. To solve this problem we make a slight modification on the fidelity and similarity functionals and thus on the Tikhonov functional.

Let S and KL be the functionals in example 3.2 and let $\sigma > 0$ be a shift Parameter. We will from now on consider the shifted fidelity functional and shifted similarity measure

$$S_\sigma : \mathcal{Z}(\Omega) \times \mathcal{P}_1(\Omega) \to \mathbb{R}$$
$$S_\sigma(z, G) := S(z + \sigma, G + \sigma \mathrm{d}x)$$

and

$$\mathrm{KL}_\sigma : \mathcal{Z}(\Omega) \times \mathcal{Z}(\Omega) \to \mathbb{R}$$
$$\mathrm{KL}_\sigma(z, w) := \mathrm{KL}(z + \sigma, w + \sigma).$$

Respectively, we modify the Tikhonov functional in (4.2) and consider the shifted Tikhonov functional

$$T_{\alpha,\sigma}^n(x) = S_\sigma(F(x), g^m) + \alpha J(x). \tag{4.5}$$

The advantage of introducing the shift parameter σ becomes clear when calculating Err for the shifted fidelity functional.

$$\mathrm{Err}_\sigma(z)(g^\dagger, g^m) := \mathrm{KL}_\sigma(z, g^\dagger) - S_\sigma(z, g^m) + S_\sigma(g^\dagger, g^m)$$

$$= \int \ln \frac{z+\sigma}{g^\dagger + \sigma} \, (\mathrm{d}g^m - \mathrm{d}g^\dagger).$$

Now Err can indeed be bounded probabilistically, since all probability densities z fulfill $z + \sigma > \sigma > 0$. We will prove a consequence of corollary (4.1) that allows this.

Corollary 4.2 *Let* $B_s(R)$ *be the ball of radius R in $H^s(\Omega)$. There exists a constant $C \geq 1$ depending only on Ω and s such that for $\rho \geq RC$ and for all $n \in \mathbb{N}$ it holds*

$$\mathbb{P}\left[\sup_{z \in B_s(R) \cap \mathcal{Z}(\Omega)} \left| \mathrm{Err}_\sigma(z)(g^\dagger, g^m) \right| \geq \frac{\rho}{\sqrt{n}} \right] \leq \exp\left(-\frac{\rho}{RC}\right).$$

Proof By the Sobolev embedding theorem and since $s > \frac{d}{2}$, it holds for $z \in B_s(R) \cap \mathcal{Z}(\Omega)$ that

$$\|z\|_\infty \leq C_1 \|z\|_{H^s} \leq C_1 R$$

for some constant $C_1 > 0$. Now we use the fact (see [21]) that for $z \in H^s(\Omega) \cap L^\infty(\Omega)$ and $\chi \in C^{\lfloor s \rfloor + 1}(\mathbb{R})$ it is $\chi \circ z \in H^s(\Omega)$ and

$$\|\chi \circ z\|_{H^s} \leq C_2 \|\chi\|_{C^{\lfloor s \rfloor + 1}} \|z\|_{H^s} \tag{4.6}$$

for some other constant $C_2 > 0$. For some positive constant σ consider the function $\chi : \mathbb{R} \to \mathbb{R}$ defined as

$$\chi(t) := \ln(t + \sigma)$$

for $t \in [0, C_1 R]$ and extended to $\mathbb{R}$ in such a way that $\chi \in C^{\lfloor s \rfloor + 1}(\mathbb{R})$. Then it holds

$$\chi \circ z \in H^s(\Omega) \qquad \text{and} \qquad \chi \circ z = \ln(z + \sigma)$$

since $0 \leq z \leq C_1 R$ almost everywhere. Now by (4.6) and setting $C_3 := \|\chi\|_{C^{\lfloor s \rfloor + 1}}$ it holds

$$\|\ln(z + \sigma)\|_{H^s} \leq C_2 C_3 R.$$

In the same way we can argue

$$\| \ln(g^{\dagger} + \sigma) \|_{H^s} \leq C_2 C_3 R.$$

We notice that

$$\left\| \ln \frac{z + \sigma}{g^{\dagger} + \sigma} \right\|_{H^s} \leq \| \ln(z + \sigma) \|_{H^s} + \| \ln(g^{\dagger} + \sigma) \|_{H^s} \leq 2 C_2 C_3 R$$

and

$$\sup_{z \in B_s(R) \cap \mathcal{Z}(\Omega)} \left| \int \ln \frac{z + \sigma}{g^{\dagger} + \sigma} \, (dg^m - g^{\dagger} dx) \right| \leq \sup_{u \in B_s(2 C_2 C_3 R)} \left| \int u (dg^m - g^{\dagger} dx) \right|$$

and thus the left hand side is even less likely to trespass a probabilistic upper bound on the right hand side: using corollary 4.1 we obtain

$$\mathbb{P} \left[\sup_{u \in B_s(2 C_2 C_3 R)} \left| \int u (dg^m - g^{\dagger} dx) \right| \geq \frac{\rho}{\sqrt{n}} \right] \leq \exp \left(-\frac{\rho}{2 C_2 C_3 R C_c} \right)$$

$$\implies \mathbb{P} \left[\sup_{z \in B_s(R) \cap \mathcal{Z}(\Omega)} \left| \int \ln \frac{z + \sigma}{g^{\dagger} + \sigma} \, (dg^m - g^{\dagger} dx) \right| \geq \frac{\rho}{\sqrt{n}} \right] \leq \exp \left(-\frac{\rho}{2 C_2 C_3 R C_c} \right).$$

Defining $C := 2 C_2 C_3 C_c$ and inserting the definition of Err_σ we finally see

$$\mathbb{P} \left[\sup_{z \in B_s(R) \cap \mathcal{Z}(\Omega)} \left| \mathrm{Err}_\sigma(z)(g^{\dagger}, g^m) \right| \geq \frac{\rho}{\sqrt{n}} \right] \leq \exp \left(-\frac{\rho}{RC} \right).$$

$\square$

With this result we are ready to study convergence rates.

4.3   Probabilistic Convergence Rates

Our task now is using theorem 3.1 without forgetting the peculiarity of our assumption about the measurement error, namely it is probabilistic. In fact, by corollary 4.2 the error bound depends on the number n of observations and the probability of the bound is high in the sense that a linear increase in the bound implies an exponential decay of the probability of trespassing it.

Lemma 4.1 *Let x_α^n be a minimizer of the Tikhonov functional in (4.5) with n denoting the number of observations. Let $x^\dagger$ be a J−minimizing solution of $F(x) = g^\dagger$ and let it fulfill a variational source condition of the form*

$$\beta \mathcal{D}_J^\xi(x, x^\dagger) \leq J(x) - J(x^\dagger) + \phi\left(\mathrm{KL}_\sigma(F(x), g^\dagger)\right) \qquad \text{for all } x \in X.$$

with ϕ, ξ and β as in assumption 3.3 and some $r > 0$. Let R as in (4.3), C as in corollary 4.2 and $\rho \geq CR$. Then for all $\alpha > 0$ we have

$$\mathbb{P}\left[\beta \mathcal{D}_J^\xi(x_\alpha^n, x^\dagger) \leq \frac{\rho}{\alpha\sqrt{n}} + (-\phi)^*\left(-\frac{1}{\alpha}\right)\right] \geq 1 - \exp\left(-\frac{\rho}{RC}\right).$$

Let α be optimally chosen, i.e. such that

$$\frac{1}{\alpha} \in \partial(-\phi)\left(\frac{\rho}{\sqrt{n}}\right).$$

Then

$$\mathbb{P}\left[\beta \mathcal{D}_J^\xi(x_\alpha^n, x^\dagger) \leq \phi\left(\frac{\rho}{\sqrt{n}}\right)\right] \geq 1 - \exp\left(-\frac{\rho}{RC}\right)$$

and by the concavity of ϕ

$$\mathbb{P}\left[\mathcal{D}_J^\xi(x_\alpha^n, x^\dagger) \leq \rho\,\phi\left(\frac{1}{\sqrt{n}}\right)\right] \geq 1 - \exp\left(-\frac{\rho}{RC}\right).$$

Proof Recall that the range of F is contained in $\mathrm{B}_s(R) \cap \mathcal{Z}(\Omega)$. The claim follows directly from theorem 3.1 with $\delta = \frac{\rho}{\sqrt{n}}$ in the event that

$$\left\{ \sup_{z \in \mathrm{B}_s(R) \cap \mathcal{Z}(\Omega)} \left|\mathrm{Err}_\sigma(z)(g^\dagger, g^m)\right| \leq \frac{\rho}{\sqrt{n}} \right\},$$

whose probability is estimated by corollary 4.2.

We end this section with a similar result that gives us the convergence rate of the regularized solutions in expectation.

Theorem 4.2 *Let x_α^n be a minimizer of the Tikhonov functional in (4.5) with n denoting the number of observations. Let $x^\dagger$ be a J-minimizing solution of $F(x) = g^\dagger$ and let it fulfill a variational source condition of the form*

$$\beta \mathcal{D}_J^\xi(x, x^\dagger) \le J(x) - J(x^\dagger) + \phi\left(\mathrm{KL}_\sigma(F(x), g^\dagger)\right) \text{ for all } x \in \mathcal{D}(F)$$

with ϕ, ξ and β as in assumption 3.3. Let α be chosen such that

$$-\frac{1}{\alpha} \in \partial(-\phi)\left(\frac{a}{b\sqrt{n}}\right),$$

where

$$a := \frac{RC}{\beta}\left(\sum_{k=1}^\infty \exp(-(k-1))k\right) \quad \text{and} \quad b := \frac{1}{\beta}\left(\sum_{k=1}^\infty \exp(-(k-1))\right)$$

Then

$$\mathbb{E}\left[\mathcal{D}_J^\xi(x_\alpha^n, x^\dagger)\right] \sim \phi\left(\frac{a}{\sqrt{n}}\right).$$

Proof Consider the events

$$E_0 := \emptyset, \qquad E_k := \left\{\sup_{z \in B_s(R) \cap \mathcal{Z}(\Omega)} \left|\mathrm{Err}_\sigma(z)(g^\dagger, g^m)\right| \le \frac{\rho_k}{\sqrt{n}}\right\},$$

where $\rho_k := kRC$ with R, C as in lemma 4.1 and $k \in \mathbb{N}$. Then by corollary 4.2 it is

$$\mathbb{P}\left[E_k\right] \ge 1 - \exp\left(-\frac{\rho_k}{RC}\right)$$
$$= 1 - \exp(-k).$$

On the event E_k we know from theorem 3.1 with $\delta = \frac{\rho_k}{\sqrt{n}}$ that for all α it holds

$$\mathcal{D}_J^\xi(x_\alpha^n, x^\dagger) \le \frac{1}{\beta}\left(\frac{\rho_k}{\alpha\sqrt{n}} + (-\phi)^*\left(-\frac{1}{\alpha}\right)\right),$$

thus we obtain

$$\mathbb{E}\left[\mathcal{D}_J^\xi(x_\alpha^n, x^\dagger)\right] = \sum_{k=1}^\infty \mathbb{P}[E_k - E_{k-1}]\mathbb{E}\left[\mathcal{D}_J^\xi(x_\alpha^n, x^\dagger)\Big|E_k - E_{k-1}\right]$$

$$\leq \sum_{k=1}^\infty \mathbb{P}[E_k - E_{k-1}]\max_{E_k}\mathcal{D}_J^\xi(x_\alpha^n, x^\dagger)$$

$$\leq \sum_{k=1}^\infty \mathbb{P}[E_k - E_{k-1}]\frac{1}{\beta}\left(\frac{\rho_k}{\alpha\sqrt{n}} + (-\phi)^*\left(-\frac{1}{\alpha}\right)\right)$$

$$\leq \sum_{k=1}^\infty \mathbb{P}[E_{k-1}^C]\frac{1}{\beta}\left(\frac{\rho_k}{\alpha\sqrt{n}}\right) + \sum_{k=1}^\infty \mathbb{P}[E_{k-1}^C]\frac{1}{\beta}(-\phi)^*\left(-\frac{1}{\alpha}\right)$$

$$\leq \frac{1}{\alpha\sqrt{n}}\frac{RC}{\beta}\underbrace{\left(\sum_{k=1}^\infty \exp(-(k-1))k\right)}_{:=a}$$

$$+ \frac{1}{\beta}\underbrace{\left(\sum_{k=1}^\infty \exp(-(k-1))\right)}_{:=b}(-\phi)^*\left(-\frac{1}{\alpha}\right),$$

where both sums converge. At this point we want to minimize the right-hand side in α to find the optimal rate of convergence and the corresponding parameter choice. But the calculation is analogous to the one in the proof of theorem 3.1, namely we minimize

$$\frac{1}{\alpha\sqrt{n}}a + b(-\phi)^*\left(-\frac{1}{\alpha}\right)$$

and obtain the minimal value

$$b\phi\left(\frac{a}{b\sqrt{n}}\right) \lesssim \phi\left(\frac{1}{\sqrt{n}}\right),$$

using concavity and $\phi(0) = 0$, where the optimal choice is given by

$$-\frac{1}{\alpha} \in \partial(-\phi)\left(\frac{a}{b\sqrt{n}}\right).$$

$\square$

Application: Parameter Identification of SDEs

5

In the publication of F.Dunker and T.Hohage we have studied, we do not only find the general convergence rate result we proved in theorem 4.2, but also an application of the theory to a specific ill-posed problem, namely parameter identification of the stationary version of an SDE. We want to examine this problem and to extend the result to a related situation in which we do take time into account. First let us give some context to the situation with an interesting problem motivated by research in biology.

5.1 Motivation

Consider a cell on an embryo in devolopement moving according to a combination of certain forces that can be partly modeled deterministically and partly randomly. The deterministic forces are assumed to be unknown and, in fact, are the object of study. Our goal is to reconstruct them based on a sample of measures of the position of the cell. To model the situation we can understand the embryo as a Riemannian manifold, whose metric is time-dependent, since the embryo developes in time. On the other hand the position of the cell at every time can be modeled by a stochastic differential equation, since the motion is governed by certain deterministic forces and some noise that can be modeled using Brownian motion. The stochastic differential equation at hand is

$$\mathrm{d}X_t = \sigma(t, X_t)\mathrm{d}B(t) + \mu(t, X_t)\mathrm{d}t. \tag{5.1}$$

Here, B is a Brownian motion, σ is a matrix-valued function denoting the *diffusion coefficient* of the process and μ is a vector-valued function denoting its *drift*. Both

© The Author(s), under exclusive license to Springer Fachmedien Wiesbaden GmbH, part of Springer Nature 2025
N. Uesseler, *Parameter Identification for a Stochastic Partial Differential Equation in the Nonstationary Case*, BestMasters,
https://doi.org/10.1007/978-3-658-50344-4_5

are defined on $[a, b] \times U$. The position of the particle is the random process X_t. The law of this process is closely related to a deterministic partial differential equation called the *Fokker–Planck equation*. The problem of reconstructing the unknown mechanisms lying behind the motion is essentially the problem of parameter identification of this partial differential equation. First we want to understand this relation presenting a formal derivation of the equation following the lines of [17]. The result can also be found in [6].

Let (M, g) be a d-dimensional Riemannian manifold and $(U, V, F : U \to V)$ a parametrisation of some open set $V \in M$ from an open set $U \in \mathbb{R}^d$. If the manifold is moving, we should think of g as a function of time, as well. The metric tensor in these coordinates will be denoted by G and its determinant by $|G|$. They are clearly also functions of time. It could also be helpful to embed the Riemannian manifold isometrically in some Euclidean space. The embedding $\Phi_t : M \to \mathbb{R}^{d+m}$ would in fact be time-dependent. The objects living on the manifold will be denoted by tildes in contrast to their corresponding versions written in coordinates. Consider now a diffusion process $\tilde{X}_t$ on the manifold with coordinates

$$\mathrm{d}X_t = d(x_1, ..., x_n)_t = \sigma(t, X_t)\mathrm{d}B(t) + \mu(t, X_t)\mathrm{d}t.$$

We will assume that the transition probabilities $\tilde{P}_{s,x}(t)$ of the diffusion process (recall this is the law of the process at time t, given x was the position at time s) have densities with respect to the (time dependent) volume measure of the manifold $V_t(q)$, i.e.

$$\tilde{P}_{s,x}(t) = \tilde{p}_{s,x}(t, .)V_t.$$

We are specially interested in the density function $\tilde{p}_{s_0,x_0}(t, q)$, where s_0 and x_0 are some fixed starting time and starting position. We will understand this as a function of t and q and will sometimes refer to it as $\tilde{f}(t, q)$.

The process X_t is a Markov process and therefore, it satisfies the *Chapman–Kolmogorov equation*, which in this setting takes the form

$$\tilde{p}_{s_0,x_0}(t + \epsilon, q) = \int_M \tilde{p}_{t,z}(t + \epsilon, q)\tilde{p}_{s_0,x_0}(t, z)\mathrm{d}V_t(z),$$

where ϵ is some small positive number. Due to its nature, the Chapman–Kolmogorov equation demands information about the whole manifold. Since we want to work locally this integral over the manifold M would be problematic. For this reason we will assume that the set V of our parametrisation covers, up to a zero-measure set, the whole manifold. By doing this we get the same equation but can replace the integration set M by the set V whenever we need to do that.

Before we can start with the derivation of the Fokker–Planck equation on our moving manifold, we recall the following facts from [17, Thm. 10.8.9] about the *drift* μ and the *diffusion* σ:

$$\mu(t, x) = \lim_{\epsilon \to 0} \frac{1}{\epsilon} \int_{|y-x|<c} (y - x) p_{t,x}(t + \epsilon, y)|G(t + \epsilon, y)|^{1/2} dy$$

$$\sigma\sigma^T(t, x) = \lim_{\epsilon \to 0} \frac{1}{\epsilon} \int_{|y-x|<c} (y - x)(y - x)^T p_{t,x}(t + \epsilon, y)|G(t + \epsilon, y)|^{1/2} dy,$$

where c is some small positive number that is not essential to the definition.

With this in mind we can start with the derivation of the equation we are seeking. Consider a function $\tilde{\xi}$ defined on the manifold, which has compact support in V. We now define the object

$$\Theta_V(t) = \int_V \tilde{\xi}(q) \tilde{p}_{s_0,x_0}(t, q) dV_t(q)$$

and its version in coordinates

$$\Theta_U(t) = \int_U \xi(x) p_{s_0,x_0}(t, x)|G(t, x)|^{1/2} dx.$$

Putting the last definitions, the Chapman–Kolmogorov equation and the theorem of Fubini together, as well as the Taylor expansion

$$\xi(x) = \xi(y) + D_y\xi(x - y) + \frac{1}{2}(x - y)^T D_y^2\xi(x - y) + R_y(x)$$

we get the following equation:

$$\Theta_U(t+\epsilon) = \Theta_V(t+\epsilon)$$

$$= \int_V \tilde{\xi}(q)\tilde{p}_{s_0,x_0}(t+\epsilon,q)\,\mathrm{d}V_{t+\epsilon}(q)$$

$$= \int_M \tilde{\xi}(q)\tilde{p}_{s_0,x_0}(t+\epsilon,q)\,\mathrm{d}V_{t+\epsilon}(q)$$

$$= \int_M \tilde{\xi}(q)\int_M \tilde{p}_{t,z}(t+\epsilon,q)\tilde{p}_{s_0,x_0}(t,z)\,\mathrm{d}V_t(z)\,\mathrm{d}V_{t+\epsilon}(q)$$

$$= \int_M \int_M \tilde{\xi}(q)\tilde{p}_{t,z}(t+\epsilon,q)\,\mathrm{d}V_{t+\epsilon}(q)\;\tilde{p}_{s_0,x_0}(t,z)\,\mathrm{d}V_t(z)$$

$$= \int_U \int_U \xi(x)p_{t,y}(t+\epsilon,x)|G(t+\epsilon,x)|^{1/2}\mathrm{d}x\;p_{s_0,x_0}(t,y)|G(t,y)|^{1/2}\mathrm{d}y$$

$$\approx \int_U \int_U \left[\xi(y) + D_y\xi(x-y) + \frac{1}{2}(x-y)^T D_y^2\xi(x-y)\right]$$
$$p_{t,y}(t+\epsilon,x)|G(t+\epsilon,x)|^{1/2}\,\mathrm{d}x\;p_{s_0,x_0}(t,y)|G(t,y)|^{1/2}\,\mathrm{d}y.$$

Dividing the equation by ϵ and taking the first summand in the integral to the left we get

$$\frac{\Theta_U(t+\epsilon) - \Theta_U(t)}{\epsilon} \approx \int_U \frac{1}{\epsilon}\int_U \left[D_y\xi(x-y) + \frac{1}{2}(x-y)^T D_y^2\xi(x-y)\right]$$
$$p_{t,y}(t+\epsilon,x)|G(t+\epsilon,x)|^{1/2}\,\mathrm{d}x\;p_{s_0,x_0}(t,y)|G(t,y)|^{1/2}\,\mathrm{d}y.$$

Letting ϵ go to zero and using the equations on drift and diffusion, we obtain

$$\Theta_U'(t) \approx \int_U \left[\sum_i \partial_i \xi(y)\mu_i(t,y)\right.$$
$$\left. + \frac{1}{2}\sum_{i,j} \partial_{i,j}\xi(y)(\sigma\sigma^T)_{i,j}(t,y)\right]p_{s_0,x_0}(t,y)|G(t,y)|^{1/2}\mathrm{d}y.$$

Now we use partial integration and the fact that ξ has compact support in U. We also replace $p_{s_0,x_0}(t,y)$ by $f(t,y)$ to lighten the notation:

$$\Theta_U'(t) \approx \int_U \xi(y)\left[\sum_i \partial_i\{-\mu_i(t,y)f(t,y)|G(t,y)|^{1/2}\}\right.$$
$$\left. + \frac{1}{2}\sum_{i,j} \partial_{i,j}\{(\sigma\sigma^T)_{i,j}(t,y)f(t,y)|G(t,y)|^{1/2}\}\right]\mathrm{d}y.$$

On the other hand, with the original definition of Θ_U we see as well

$$\Theta'_U(t) = \int_U \xi(y)\partial_t\Big[f(t, y)|G(t, y)|^{1/2}\Big]dx.$$

Since the function ξ is abitrary we can combine both equations and get rid of the integral. This way, we finally arrived at the statement

$$\partial_t\Big[f(t, y)|G(t, y)|^{1/2}\Big] = \sum_i \partial_i\big\{-\mu_i(t, y)f(t, y)|G(t, y)|^{1/2}\big\}$$
$$+ \frac{1}{2}\sum_{i,j} \partial_{i,j}\big\{(\sigma\sigma^T)_{i,j}(t, y)f(t, y)|G(t, y)|^{1/2}\big\}.$$

We will call the last equation *Fokker–Planck equation on a moving manifold.* Lastly, just for a more convenient notation, we will set

$$u(t, x) := f(t, x)|G(t, x)|^{1/2}$$

to rewrite the equation as follows:

$$\partial_t u = \sum_i \partial_i\big\{-\mu_i u\big\} + \frac{1}{2}\sum_{i,j} \partial_{i,j}\big\{(\sigma\sigma^T)_{i,j}u\big\}.$$

With this equation at hand we will reconstruct or identify the parameter μ of the equation based on measurements of the solution. In fact, the measurements are real-isations of the underlying probability law of the random process whose probability density is the solution of the equation. The theory developed in the sections above will give us a method of reconstruction together with probabilistic error estimates depending on the number of measurements that are available.

5.2   The Stationary Case on $\mathbb{R}^d$

The general Fokker–Planck equation is given as

$$\partial_t u = \sum_i \partial_i\big\{-\mu_i u\big\} + \frac{1}{2}\sum_{i,j} \partial_{i,j}\big\{(\sigma\sigma^T)_{i,j}u\big\}$$

or equivalently as

$$\partial_t u = \nabla \cdot (-\mu u) + \frac{1}{2}\nabla^2 : (u\,\sigma\sigma^T).$$

We will consider the case of a constant diffusion which reduces the equation to the form

$$\partial_t u = \nabla \cdot (-\mu u) + \frac{1}{2}\nabla \cdot (\sigma\sigma^T \nabla u)$$

and if the solution u does not change in time the left-hand side vanishes. More precisely we have the equation

$$0 = \nabla \cdot (-\mu u) + \frac{1}{2}\nabla \cdot (\sigma\sigma^T \nabla u)$$

with homogeneous Neumann boundary conditions

$$u(\mu \cdot \nu) = 0 = \sigma\sigma^T \nabla u \cdot \nu \qquad \text{on } \partial\Omega \text{ for } \nu \text{ the outward normal vector.}$$

This is the case that F.Dunker and T.Hohage studied. We will consider the corresponding weak elliptic problem: Find $u \in H^1(\Omega)$, such that

$$\int u \, dx = 1, \qquad B_\mu(u, v) = 0$$

for all $v \in H^1(\Omega)$, where

$$B_\mu(u, v) := \int \left(-\mu u \cdot \nabla v + \frac{1}{2}\sigma\sigma^T \nabla u \cdot \nabla v\right) dx.$$

Solutions to the Fokker–Planck equation tend to the solution of this problem as $t \to \infty$, hence u is called the stationary solution.

Let $F : L^\infty(\Omega, \mathbb{R}^d) \to L^2(\Omega)$ be defined as $F(\mu) = u^\dagger$, where $u^\dagger$ solves the problem above. In order to estimate the drift coefficient μ based on observations of particle positions that obey the probability distribution corresponding to $u^\dagger$, we want to apply theorem 4.2. For that we need to ensure that the assumptions are satisfied. Let us start with a basic fact about the operator.

Theorem 5.1   *$F : L^\infty(\Omega, \mathbb{R}^d) \to L^2(\Omega)$ is Fréchet-differentiable and the derivative $F'(\mu)[h] =: u_\mu^h \in H_\diamond^1(\Omega) := \{u \in H^1(\Omega) : \int u \, \mathrm{d}x = 0\}$ solves the variational problem*

$$B_\mu(u_\mu^h, v) = \int F(\mu) \, h \cdot \nabla v \, \mathrm{d}x, \qquad \text{for all } v \in H_\diamond^1(\Omega).$$

Moreover, F fulfills a tangential cone condition:

$$\|F(\mu + h) - F(\mu) - F'(\mu)[h]\|_{L^2(\Omega)} \le C_\mu \|h\|_\infty \|F(\mu + h) - F(\mu)\|_{L^2(\Omega)}$$

for all $\mu, h \in L^\infty(\Omega)$, where $C_\mu > 0$ is the norm of L_μ^{-1}, where $L_\mu : H_\diamond^1(\Omega) \to H_\diamond^1(\Omega)^\star$ is the operator associated to $B_\mu|_{H_\diamond^1(\Omega) \times H_\diamond^1(\Omega)}$.

Proof Let us define $\tilde{u} := F(\mu + h) - F(\mu)$. Since the range of F consists of probability densities, $\tilde{u}$ is in $H_\diamond^1(\Omega)$. For $v \in H_\diamond^1(\Omega)$, it holds

$$\begin{aligned}
B_\mu(\tilde{u}, v) &= \int \left(-\mu \tilde{u} \cdot \nabla v + \frac{1}{2} \sigma \sigma^T \nabla \tilde{u} \cdot \nabla v \right) \mathrm{d}x \\
&= \int \left(-\mu \, F(\mu + h) \cdot \nabla v + \frac{1}{2} \sigma \sigma^T \, \nabla F(\mu + h) \cdot \nabla v \right) \mathrm{d}x \\
&\quad + \int \left(\mu \, F(\mu) \cdot \nabla v - \frac{1}{2} \sigma \sigma^T \, \nabla F(\mu) \cdot \nabla v \right) \mathrm{d}x \\
&= \int \left(-\mu \, F(\mu + h) \cdot \nabla v + \frac{1}{2} \sigma \sigma^T \, \nabla F(\mu + h) \cdot \nabla v \right) \mathrm{d}x - \underbrace{B_\mu(F(\mu), v)}_{=0}
\end{aligned}$$

and thus

$$\begin{aligned}
B_\mu(\tilde{u}, v) &= \int \left(-(\mu + h) \, F(\mu + h) \cdot \nabla v + \frac{1}{2} \sigma \sigma^T \, \nabla F(\mu + h) \cdot \nabla v \right) \mathrm{d}x \\
&\quad + \int h \, F(\mu + h) \cdot \nabla v \, \mathrm{d}x \\
&= \underbrace{B_{\mu+h}(F(\mu + h), v)}_{=0} + \int h \, F(\mu + h) \cdot \nabla v \, \mathrm{d}x \\
&= \int h \, (F(\mu) + \tilde{u}) \cdot \nabla v \, \mathrm{d}x.
\end{aligned}$$

And we can bound the right-hand side in the following way:

$$\int h\,(F(\mu)+\tilde{u})\cdot\nabla v\,\mathrm{d}x \le \|v\|_{H^1(\Omega)}\|F(\mu)+\tilde{u}\|_{L^2(\Omega)}\|h\|_\infty$$

$$\le \|v\|_{H^1(\Omega)}(\|F(\mu)\|_{L^2(\Omega)}+\|\tilde{u}\|_{H^1(\Omega)})\|h\|_\infty$$

and thus for all $v \ne 0$ in $H^1_\diamond(\Omega)$

$$\left\langle L_\mu\tilde{u},\,\frac{v}{\|v\|_{H^1(\Omega)}}\right\rangle = B_\mu\left(\tilde{u},\,\frac{v}{\|v\|_{H^1(\Omega)}}\right)$$

$$\le (\|F(\mu)\|_{L^2(\Omega)}+\|\tilde{u}\|_{H^1(\Omega)})\|h\|_\infty$$

which implies

$$\|L_\mu\tilde{u}\|_{H^1_\diamond(\Omega)^\star} \le (\|F(\mu)\|_{L^2(\Omega)}+\|\tilde{u}\|_{H^1(\Omega)})\|h\|_\infty.$$

So, we obtain

$$\|\tilde{u}\|_{H^1(\Omega)} = \|L_\mu^{-1}L_\mu\tilde{u}\|_{H^1(\Omega)}$$

$$\le \|L_\mu^{-1}\|\|L_\mu\tilde{u}\|_{H^1(\Omega)}$$

$$\le C_\mu(\|F(\mu)\|_{L^2(\Omega)}+\|\tilde{u}\|_{H^1(\Omega)})\|h\|_\infty$$

and thus

$$(1-C_\mu\|h\|_\infty)\|\tilde{u}\|_{H^1(\Omega)} \le C_\mu\|h\|_\infty\|F(\mu)\|_{L^2(\Omega)}.$$

Since the left-hand side converges to $\|\tilde{u}\|_{H^1(\Omega)}$ as $\|h\|_\infty \to 0$ and the right-hand side converges to 0, this implies that

$$\|F(\mu+h)-F(\mu)\|_{L^2(\Omega)} \le \|F(\mu+h)-F(\mu)\|_{H^1(\Omega)} = \|\tilde{u}\|_{H^1(\Omega)}$$

is 0 in the limit and thus, the operator F defined on L^∞ is continuous in the $H^1(\Omega)$-topology (and thus also in the $L^2(\Omega)$-topology). Now we use the definition of u_μ^h and consider

$$B_\mu(\tilde{u} - u_\mu^h, v) = B_\mu(\tilde{u}, v) - B_\mu(u_\mu^h, v)$$

$$= \int h\,(F(\mu) + \tilde{u}) \cdot \nabla v\,\mathrm{d}x - \int F(\mu)\,h \cdot \nabla v\,\mathrm{d}x$$

$$= \int h\,\tilde{u} \cdot \nabla v\,\mathrm{d}x.$$

As before, we can estimate the right-hand side and we obtain

$$\|\tilde{u} - u_\mu^h\|_{L^2(\Omega)} \leq \|\tilde{u} - u_\mu^h\|_{H^1(\Omega)}$$

$$\leq C_\mu \|\tilde{u}\|_{H^1(\Omega)} \|h\|_\infty,$$

which implies

$$\|F(\mu + h) - F(\mu) - u_\mu^h\|_{H^1(\Omega)} = \|\tilde{u} - u_\mu^h\|_{H^1(\Omega)}$$

$$\leq C_\mu \|\tilde{u}\|_{H^1(\Omega)} \|h\|_\infty$$

$$= C_\mu \|F(\mu + h) - F(\mu)\|_{H^1(\Omega)} \|h\|_\infty,$$

which proves the tangential cone condition in the $H^1(\Omega)$-norm. The condition for the $L^2(\Omega)$-norm follows then from the Poincaré inequality in the space $H^1_\diamond(\Omega)$. Moreover, dividing by $\|h\|_\infty$, and letting it tend to 0, together with the continuity of F at μ and the fact that the map $h \mapsto u_\mu^h$ is linear and bounded, we also obtain that F is Fréchet-differentiable and that, indeed

$$F'(\mu)[h] = u_\mu^h.$$

$\square$

The next step is to prove a lemma that relates the Kullback–Leibler divergence with the $L^2(\Omega)-$norm (see [3]).

Lemma 5.1 *for all nonnegative functions $x, y \in L^\infty(\Omega)$ with $x - y \in L^2(\Omega)$ and $y > 0$ a.e. it holds*

$$\|x - y\|^2_{L^2(\Omega)} \leq 2 \max(\|x\|_\infty, \|y\|_\infty)\mathrm{KL}(x, y).$$

Proof Consider $v \in (0, 1]$ and define the function

$$\gamma(u) = u \ln\left(\frac{u}{v}\right) - u + v - \frac{1}{2}(u - v)^2$$

for $u \in [0, 1]$. By differentiating twice one can see that the function is convex and by minimizing it, one sees that $\gamma \geq 0$. Thus defining $M := \max(\|x\|_\infty, \|y\|_\infty)$ and

$$\tilde{x}(s) := \frac{x(s)}{M}, \qquad\qquad \tilde{y}(s) := \frac{y(s)}{M}$$

we obtain

$$\frac{1}{2}(\tilde{x}(s) - \tilde{y}(s))^2 \leq \tilde{x}(s) \ln\left(\frac{\tilde{x}(s)}{\tilde{y}(s)}\right) - \tilde{x}(s) + \tilde{y}(s)$$

$$\Longleftrightarrow \frac{1}{2}(\frac{x(s)}{M} - \frac{y(s)}{M})^2 \leq \frac{x(s)}{M} \ln\left(\frac{x(s)}{y(s)}\right) - \frac{x(s)}{M} + \frac{y(s)}{M}$$

$$\Longleftrightarrow (x(s) - y(s))^2 \leq 2M \left(x(s) \ln\left(\frac{x(s)}{y(s)}\right) - x(s) + y(s)\right).$$

The claim follows from integration. $\qquad\qquad\qquad\qquad\qquad\qquad\qquad\qquad\square$

Notice that the last result implies an inequality of the same type if one replaces KL with KL_σ. At this point it will be useful to recall a regularity result in elliptic equations (for similar but not precisely the needed result in the literature, see e.g. [13, Thm. 8.13] for the case of Dirichlet-boundary conditions or [4, Them. 1.1] for the case of Neumann-boundary conditions with smooth coefficients), namely the fact that for coefficients in $C^{s-1,1}(\bar{\Omega})$ of the elliptic operator the solution u is in $H^s(\Omega)$. On the other hand, to ensure that the coefficient μ is in $C^{s-1,1}(\bar{\Omega})$, we can impose that $\mu \in H^{s'}(\Omega)$ with $s' > s + \frac{d}{2}$ (see e.g. [1, Thm. 8.13]). From these regularity results one can also read off the explicit dependencies on the $H^{s'}(\Omega)$-norm of μ. Indeed, if μ is in a bounded set in $H^{s'}(\Omega)$, the range of F is bounded in $H^s(\Omega)$.

From similar considerations we can see the compactness of $F'(\mu)$, which will be useful. Under the same regularity conditions on the coefficients, the inhomogeneous equation $L_\mu u_\mu^h = \nabla \cdot (h F(\mu))$ still ensures H^s-regularity of u_μ^h as long as the right-hand side is in $H^{s-2}(\Omega)$. Indeed, if we choose $s > \frac{d}{2}$, then we can use that $H^s(\Omega)$ is an algebra and the product $h F(\mu)$ (recall $s' > s$) is again in $H^s(\Omega)$ and its divergence in H^{s-1}, which is sufficient. This implies then that $F'(\mu)[h]$ is in $H^s(\Omega)$, which embeds compactly in $L^2(\Omega)$. Thus the operator $F'(\mu) : L^\infty(\Omega) \to L^2(\Omega)$ is compact.

With these facts we can now show that theorem 4.2 is in fact applicable to our situation, namely that the variational source condition is fulfilled by the ground truth $\mu^\dagger$, for which it holds $F(\mu^\dagger) = u^\dagger$. Restricting the domain of F to a bounded set $\mathcal{B} \subset H^{s'}(\Omega)$ containing the ground truth we guarantee the range of F to be also bounded in $H^s(\Omega)$. Before stating the thorem let us see another useful fact derived by B.Hofmann and P.Mathé in [20, Cor. 2]. It states basically that in Hilbert spaces, general source conditions can always be fulfilled.

Theorem 5.2 *Let $K : X \to X$ be a compact, self-adjoint, injective and nonnegative linear operator. Then for every $x \in X$ there is a concave index function ϕ (as in assumption 3.3) such that $x = \phi(K)w$ for some $w \in X$.*

This theorem together with the fact that general source conditions imply variational source conditions [11, Thm. 3.1] by J.Flemming and B.Hofmann was used in [7, Prop. 11] to obtain the desired result, whose proof we present. Here we notice that an additional argument should be necessary to apply the theorem to the operator $F'(\mu^\dagger)^* F'(\mu^\dagger)$, since $F'(\mu^\dagger)$ is not injective.

Theorem 5.3 *Let $J(\mu) = \|\mu\|^2_{H^{s'}(\Omega)}$ be the penalty term with $s' > s + \frac{d}{2}$ and $s > \frac{d}{2}$, then every $\mu^\dagger \in H^{s'}(\Omega, \mathbb{R}^d)$ satisfies a variational source condition of the form*

$$\beta \mathcal{D}_J^\xi(\mu, \mu^\dagger) \le J(\mu) - J(\mu^\dagger) + \phi\left(\mathrm{KL}_\sigma(F(\mu), u^\dagger)\right)$$

in some $H^{s'}-$ball.

Proof By last theorem (or a modified version that does not require injectivity of $F'(\mu^\dagger)^* F'(\mu^\dagger)$) there is some index function Θ such that

$$\mu^\dagger = \Theta(F'(\mu^\dagger)^* F'(\mu^\dagger))w$$

for some $w \in H^{s'}(\Omega, \mathbb{R}^d)$. This implies that for some other index function λ, $\mu^\dagger$ also satisfies a variational source condition of the type

$$\beta \mathcal{D}_J^\xi(\mu, \mu^\dagger) \le J(\mu) - J(\mu^\dagger) + \lambda\left(\|F'(\mu)[\mu - \mu^\dagger]\|^2_{L^2(\Omega)}\right).$$

Now recall the tangential cone condition that F satisfies

$$\|F(\mu + h) - F(\mu) - F'(\mu)[h]\|_{L^2(\Omega)} \le C_\mu \|h\|_\infty \|F(\mu + h) - F(\mu)\|_{L^2(\Omega)},$$

which implies

$$\|F'(\mu)[\mu - \mu^\dagger]\|_{L^2(\Omega)} \le C_\mu \|\mu - \mu^\dagger\|_\infty \|F(\mu^\dagger) - F(\mu)\|_{L^2(\Omega)} + \|F(\mu^\dagger) - F(\mu)\|_{L^2(\Omega)}.$$

Thus for all $\mu \in H^{s'}(\Omega, \mathbb{R}^d)$ with $C_\mu \|\mu - \mu^\dagger\|_\infty \le 1$, the variational source condition becomes

$$\beta \mathcal{D}_J^\xi(\mu, \mu^\dagger) \le J(\mu) - J(\mu^\dagger) + \lambda \left(4\|F(\mu^\dagger) - F(\mu)\|_{L^2(\Omega)}^2 \right)$$

and with the monotonicity of λ and the relation between the Kullback–Leibler divergence and the $L^2(\Omega)$−norm

$$\beta \mathcal{D}_J^\xi(\mu, \mu^\dagger) \le J(\mu) - J(\mu^\dagger) + \lambda \left(8 \max(\|u^\dagger\|_\infty, \|F(\mu)\|_\infty) \mathrm{KL}_\sigma(F(\mu), u^\dagger) \right).$$

Finally, since the range of F is bounded in $H^s(\Omega)$ by the assumption on s' and this space embeds continuously in $L^\infty(\Omega)$, we can bound the term $\max(\|u^\dagger\|_\infty, \|F(\mu)\|_\infty)$ by a constant. By choosing another index function ϕ we get

$$\beta \mathcal{D}_J^\xi(\mu, \mu^\dagger) \le J(\mu) - J(\mu^\dagger) + \phi \left(\mathrm{KL}_\sigma(F(\mu), u^\dagger) \right).$$

$\square$

Hence we can apply the theory to this particular problem.

5.3   The Time-Dependent Case on $\mathbb{R}^d$

We are now interested in the time-dependent problem. In this setting there are n particles moving independently according to the SDE (5.1) introduced in the motivation. In a given interval $[0, T]$ of time, we are allowed to measure the position of the particles at times $0 = t_0 < t_1 < ... < t_M = T$. Based on this information we want to estimate the drift μ in the SDE using the Fokker–Planck equation and the log-likelihood functional as before.

5.3.1 The Forward Operator and the Fidelity Term

Let us consider the process in a bounded domain $\Omega \subset \mathbb{R}^d$. The complete data would consist of n positions at $M + 1$ times. We denote the position of the particle j at times $t_0, \dots, t_M$ by $q_j = (q_j^0, \dots, q_j^M) \in \Omega^{M+1}$, which is a realization of the random variable $Q_j = (Q_j^0, \dots, Q_j^M)$. To use the log-likelihood approach we need to find the density function (for example with respect to the Lebesgue measure) of the law of Q_j, which is the joint law of $Q_j^0, \dots, Q_j^M$. Then we could insert the realizations q_j in the density function to get the value of the likelihood. Notice that the random variables $Q_0, \dots, Q_n$ are independent random variables since the particles do not affect each other by construction of the experiment.

Let us define for fixed j the probability measure on Ω^{M+1}

$$\nu(\mathrm{d}x^0, \dots, \mathrm{d}x^M)$$

as the joint law of $Q_j^0, \dots, Q_j^M$. To calculate the density of ν w.r.t. the Lebesgue measure we use the disintegration theorem on ν in the correct order, namely we disintegrate progressively backwards in time:

$$\nu(\mathrm{d}x^0, \dots, \mathrm{d}x^M) = \nu_{x^0,\dots,x^{M-1}}(\mathrm{d}x^M)\hat{\nu}_{0,\dots,M-1}(\mathrm{d}x^0, \dots, \mathrm{d}x^{M-1}),$$

where $\hat{\nu}_{0,\dots,M-1}$ refers to the marginal distribution of ν on the components given as subscript, i.e. if $\pi_{0,\dots,M-1}$ is the projection of Ω^{M+1} into the first $0, \dots, M-1$ components, then $\hat{\nu}_{0,\dots,M-1} = (\pi_{0,\dots,M-1})_*\nu$ is the pushforward of ν under $\pi_{0,\dots,M-1}$. It is the joint law of $Q_j^0, \dots Q_j^{M-1}$. By using disintegration successively on the marginal distribution we can write

$$\nu(\mathrm{d}x^0, \dots \mathrm{d}x^M)$$
$$= \nu_{x^0,\dots,x^{M-1}}(\mathrm{d}x^M)\nu_{x^0,\dots,x^{M-2}}(\mathrm{d}x^{M-1}) \cdots \nu_{x^0}(\mathrm{d}x^1)\hat{\nu}_0(\mathrm{d}x^0),$$

where $\hat{\nu}_0$ is the law of Q_j^0, which we assume to have the density $g^0 > 0$ with respect to the Lebesgue measure on Ω. Now by virtue of the Markov property of the process, we obtain

$$\nu(\mathrm{d}x^0, \dots \mathrm{d}x^M)$$
$$= \nu_{x^{M-1}}(\mathrm{d}x^M)\nu_{x^{M-2}}(\mathrm{d}x^{M-1}) \cdots \nu_{x^0}(\mathrm{d}x^1)g^0\mathrm{d}x^0.$$

The measure $\nu_{x^{i-1}}(\mathrm{d}x^i)$ describes the law of Q^i_j given $Q^{i-1}_j = x^{i-1}$. Its density is given by the solution of the Fokker–Planck equation corresponding to the SDE (5.1), namely (for simplicity, we now assume a scalar diffusion coefficient)

$$\partial_t u = \nabla \cdot (-\mu u) + \frac{1}{2}\nabla \cdot (\frac{\sigma^2}{2}\nabla u) \qquad \text{in } [0, T] \times \Omega,$$

with total mass preserving Neumann boundary conditions

$$u(\mu \cdot v) = 0 = \nabla u \cdot v \quad \text{on } \partial\Omega \text{ for } v \text{ the outward normal vector.}$$

for the initial condition that $u(t_{i-1}) = \delta_{x^{i-1}}$ is a Dirac measure in x^{i-1}. This corresponds to the definition of the Green's function $g(t_i, x^i; t_{i-1}, x^{i-1})$ of the Fokker–Planck equation, that is, $u(t_i, x^i) = g(t_i, x^i; t_{i-1}, x^{i-1})$.

Thus we can write

$$\nu(\mathrm{d}x^0, ... \mathrm{d}x^M) = g^0(x^0) \prod_{i=1}^{M} g(t_i, x^i; t_{i-1}, x^{i-1})\mathrm{d}x^0... \mathrm{d}x^M.$$

Having the density of ν at hand, we can define the likelihood and the log-likelihood of the data $q = (q_1, ..., q_n)$ with respect to the Lebesgue measure on $\Omega^{(M+1)n}$, namely

$$S_L(q) = \prod_{j=1}^{n} g^0(q^0_j) \prod_{i=1}^{M} g(t_i, q^i_j; t_{i-1}, q^{i-1}_j)$$

$$S_l(q) = \sum_{j=1}^{n}\left[\ln g^0(q^0_j) + \sum_{i=1}^{M} \ln g(t_i, q^i_j; t_{i-1}, q^{i-1}_j)\right].$$

Now we define the forward operator as follows

$$F : L^\infty(\Omega; \mathbb{R}^d) \longrightarrow (L^2(\Omega \times \Omega))^M, \qquad \mu \longmapsto \begin{bmatrix} g_\mu(t_1, \cdot; t_0, \cdot) \\ g_\mu(t_2, \cdot; t_1, \cdot) \\ \vdots \\ g_\mu(t_M, \cdot; t_{M-1}, \cdot) \end{bmatrix}.$$

With this definition of F, we can define the fidelity functional $S(F(\mu), q)$ that will be responsible for comparing reconstructions with observations. Let $g^0 > 0$

be given, that is, we know the probability density of the starting positions of the particles. Let us define the set $\mathcal{MZ}(\Omega)$ of bistochastic Markov kernels from Ω to Ω which are absolutely continuous with respect to the Lebesgue measure, e.g. $F_i(\mu)$ is in that set, since for every $x,\, y \in \Omega$, the maps $F_i(\mu)(\cdot, x)$ and $F_i(\mu)(y, \cdot)$ are probability densities on Ω. Moreover, let $\mathcal{P}(\Omega^{M+1})$ be the set of probability measures on Ω^{M+1} and notice that $\mathcal{MZ}(\Omega)^M$ can be understood as a subset of $\mathcal{P}(\Omega^{M+1})$, since for $(z_1, \ldots, z_M)^T \in \mathcal{MZ}(\Omega)^M$ the product $z^\pi := g^0 \prod_{i=1}^M z_i$ is a probability density on Ω^{M+1} with respect to the Lebesgue measure. Define along the lines of example 3.2 the empirical probability measure on Ω^{M+1} associated to the data $q = (q_1, \ldots, q_n)$,

$$G^n = \frac{1}{n} \sum_{i=1}^n \delta_{q_j}$$

and

$$S : \mathcal{MZ}(\Omega)^M \times \mathcal{P}(\Omega^{M+1}) \to \mathbb{R}, \quad \left(\begin{pmatrix} z_1 \\ \vdots \\ z_M \end{pmatrix}, G \right) \mapsto \begin{cases} -\int \ln z^\pi \, \mathrm{d}G & z^\pi > 0 \ G\text{-a.e.} \\ \infty & else. \end{cases}$$

Let us now recall the Kullback–Leibler divergence $\mathrm{KL}(z, u) = \int u \ln \frac{u}{z} + z - u \, \mathrm{d}x$ and consider its following version for this setting

$$\mathrm{KL}^\pi : \mathcal{MZ}(\Omega)^M \times \mathcal{MZ}(\Omega)^M \to \mathbb{R}$$

$$\left(\begin{pmatrix} z_1 \\ \vdots \\ z_M \end{pmatrix}, \begin{pmatrix} u_1 \\ \vdots \\ u_M \end{pmatrix} \right) \mapsto \int (u^\pi \ln \frac{u^\pi}{z^\pi} + z^\pi - u^\pi) \, \mathrm{d}x.$$

After having defined these functionals we are also able to give the error functional to measure the error between the exact data $g^\dagger \in \mathcal{MZ}(\Omega)^M$ and the measurement $G^n \in \mathcal{P}(\Omega^{M+1})$.

$$\mathrm{Err}(z)(g^\dagger, G^n) = \mathrm{KL}^\pi(z, g^\dagger) - S(z, G^n) + S(g^\dagger, G^n)$$

$$= \int \ln \frac{z^\pi}{(g^\dagger)^\pi} \, (\mathrm{d}G^n - (g^\dagger)^\pi \, \mathrm{d}x).$$

This can be controlled using the same techniques we used for the stationary case, which means that the general strategy is the same. Now we study, as before, the properties of the forward operator F.

5.3.2   A Sufficient Tangential Cone Condition

In this section we want to see the reduction of the variational source condition, which is necessary to apply the theorem on rates of convergence, to the simpler tangential cone condition

$$\left\| F(\mu^{\dagger}+h) - F(\mu^{\dagger}) - F'(\mu^{\dagger})[h] \right\|_{L^2(\Omega \times \Omega)^M}$$

$$\lesssim \|h\|_{L^\infty(\Omega; \mathbb{R}^d)} \| F(\mu^{\dagger}+h) - F(\mu^{\dagger}) \|_{L^2(\Omega \times \Omega)^M} \quad \text{for all } h \in B_R^r. \qquad (5.2)$$

Under the assumption that this property of F holds, we are almost ready to reformulate theorem 5.3 in the context of the nonstationary case. The ingridient that is missing in order to transport also the proof to the current situation is the relationship between the norm in which the tangential cone condition is given, namely the $L^2(\Omega \times \Omega)-$ norm and the functional we are using to compare the exact data with the reconstructions, namely KL^π. Given such a relationship, we will be able to formulate the variational source condition in terms of KL^π. In the elliptic case lemma 5.1 was helpful as it will be in the current case. Nontheless a direct application of lemma 5.1 would yield the relation

$$\|z^\pi - u^\pi\|^2_{L^2(\Omega^{n(M+1)})} \lesssim \text{KL}^\pi(z, u),$$

for two given reconstructions $z, u \in \left(L^2(\Omega \times \Omega) \right)^M$. We need to combine lemma 5.1 with another calculation in order to get the analogous relation in this setting. We formulate it as a lemma

Lemma 5.2 (Estimate for Kullback–Leibler divergence). *All $z, u \in \mathcal{MZ}(\Omega)^M$ satisfy*

$$\|z - u\|^2_{L^2(\Omega \times \Omega)^M} \lesssim \max(\|u^\pi + \tau\|_{L^\infty(\Omega^{M+1})}, \|z^\pi + \tau\|_{L^\infty(\Omega^{M+1})}) \text{KL}^\pi_\tau(z, u).$$

Proof We need to prove the relation

$$\|z - u\|^2_{L^2(\Omega \times \Omega)^M} \lesssim \|z^\pi - u^\pi\|^2_{L^2(\Omega^{n(M+1)})}. \qquad (5.3)$$

Then the result follows using lemma 5.1. To see that let $0 < C := \inf g_0$, where g_0 was the known distribution of the starting position. Then we see with $\int_\Omega v(x,\hat{x})\,dx = 1 = \int_\Omega v(x,\hat{x})\,d\hat{x}$ for all $v \in \mathcal{MZ}(\Omega)$ that any $z \in \mathcal{MZ}(\Omega)^M$ satisfies

$$z_i(x^i, x^{i-1}) = \int_\Omega z_{i-1}(x^{i-1}, x^{i-2}) \cdots \int_\Omega z_1(x^1, x^0)\,dx^0 \cdots dx^{i-2}$$

$$\cdot z_i(x^i, x^{i-1}) \cdot \int_\Omega z_{i+1}(x^{i+1}, x^i) \cdots \int_\Omega z_M(x^M, x^{M-1})\,dx^M \cdots dx^{i+1}$$

$$= \int_{\Omega^{M-1}} z^\pi(x^0, \ldots, x^M)/g^0(x^0)\,dx^0 \cdots dx^{i-2}\,dx^{i+1} \cdots dx^M$$

for all $i = 1, \ldots, M$. Thus, with Jensen's inequality we obtain

$$\|z_i - u_i\|^2_{L^2(\Omega\times\Omega)}$$

$$= \int_{\Omega^2} \left(\int_{\Omega^{M-1}} \frac{z^\pi(x^0, \ldots, x^M) - u^\pi(x^0, \ldots, x^M)}{g^0(x^0)}\,dx^0 \cdots dx^{i-2}\,dx^{i+1} \cdots dx^M \right)^2 dx^{i-1}\,dx^i$$

$$\leqq |\Omega|^{M-1} \int_{\Omega^{M+1}} \frac{|z^\pi(x^0, \ldots, x^M) - u^\pi(x^0, \ldots, x^M)|^2}{g^0(x^0)^2}\,dx^0 \cdots dx^M$$

$$\leqq |\Omega|^{M-1} \|z^\pi - u^\pi\|^2_{L^2(\Omega^{M+1})} / \inf_{x\in\Omega} g^0(x)^2,$$

which shows (5.3). $\qquad\qquad\square$

Finally we can state the theorem that allows us to use the theory developed in the sections above, namely the fact that the variational source condition is fulfilled under the assumption that the tangential cone condition holds and thus the probabilistic convergence rates results 4.1 and 4.2 can be applied to the ill-posed problem in the nonstationary case. The imposed regularity on the coefficients this time is in principle $C^{1,\lambda}(\Omega)$ for some $\lambda > 0$ in order to have the existence and the uniqueness of the fundamental solution of the parabolic equation (see e.g. [16, Sec. 4]). This is also the required regularity later to obtain the estimates in lemma 5.3. Alternatively we can impose Sobolev regularity and use the embedding theorems (see e.g.[1, Thm. 8.13]), thus we assume $H^r(\Omega)$ regularity for the coefficients, where $r > \frac{d}{2} + 1$.

Theorem 5.4 *Let F fulfil condition (5.2). $J(\mu) = \|\mu\|^2_{H^r}$ be the penalty term with $r > \frac{d}{2} + 1$, then every $\mu^\dagger \in H^r(\Omega, \mathbb{R}^d)$ satisfies a variational source condition of the form*

$$\beta \mathcal{D}_J^{\xi}(\mu, \mu^{\dagger}) \leq J(\mu) - J(\mu^{\dagger}) + \phi\left(\mathrm{KL}_{\sigma}^{\pi}(F(\mu), u^{\dagger})\right)$$

for all $\mu - \mu^{\dagger} \in B_R^r$.

5.3.3   A Weaker Tangential Cone Condition

In this section we will prove the validity of a tangential cone condition weaker than (5.2) in our parameter identification setting. This condition is not sufficient to use the results of the sections above but takes us a step further in the direction of its proof, which will be treated in future work. To state the weaker tangential cone condition we aim to prove, we first introduce the operator P.

For a given drift μ in the set

$$H_N^r = \{\mu \in H^r(\Omega; \mathbb{R}^d) \mid \mu \cdot \nu = 0 \text{ on } \partial\Omega\}$$

of vectorfields with vanishing normal component on the boundary, where $r > 1 + \frac{d}{2}$, consider the elliptic operator and its corresponding bilinear form

$$L_{\mu} : H^1(\Omega) \to H^1(\Omega)^*, \qquad L_{\mu}u = -\Delta(\tfrac{\sigma^2}{2}u) + \nabla \cdot (\mu u)$$

$$B_{\mu} : H^1(\Omega) \times H^1(\Omega) \to \mathbb{R}, \quad B_{\mu}(u, v) = \int_{\Omega} -\mu u \cdot \nabla v + \tfrac{\sigma^2}{2}\nabla u \cdot \nabla v \, dx.$$

let us fix a position x^0 and consider the Fokker–Planck equation corresponding to the SDE, namely the parabolic partial differential equation

$$
\begin{aligned}
\partial_t u + L_{\mu}u &= 0 & &\text{in } [0, T] \times \Omega, & (5.4)\\
(\tfrac{\sigma^2}{2}\nabla u - \mu u) \cdot \nu &= 0 & &\text{on } [0, T] \times \partial\Omega,\\
u &= \delta_{x^0} & &\text{at } t = 0,
\end{aligned}
$$

or in its weak formulation

$$
\begin{aligned}
\langle \partial_t u, v \rangle + B_{\mu}(u, v) &= 0 & &\text{for all } t \in (0, T], \, v \in H^1(\Omega),\\
\lim_{t \searrow 0} \langle u, v \rangle &= v(x^0) & &\text{for all } v \in C^0(\overline{\Omega}).
\end{aligned}
$$

Now we define the operator P as the operator that maps a drift $\mu \in H_N^r$ to the function $(t, x, x^0) \mapsto g_\mu(t, x, 0, x^0)$, where g_μ is, as before, the Green's function of the PDE, that is, the function $u(t, x) := g_\mu(t, x, 0, x^0)$ solves the PDE above. Thus $P(\mu)$ is the Green's function, which due to the time-independence of the coefficients in the equation only needs one argument for time. Operator P and forward operator F of section 5.3.1 are related in the following way: consider the ith component of F and notice

$$F(\mu)_i(x^i, x^{i-1}) = P(\mu)(t_i - t_{i-1}, x^i, x^{i-1}).$$

The tangential cone condition that we will prove to hold in this section is the following:

$$\left\| P(\mu^\dagger + h) - P(\mu^\dagger) - P'(\mu^\dagger)[h] \right\|_{L_\alpha^\infty L^2(\Omega \times \Omega)}$$
$$\lesssim \|h\|_{L^\infty(\Omega)} \| P(\mu^\dagger + h) - P(\mu^\dagger) \|_{L_\alpha^\infty L^2(\Omega \times \Omega)} \qquad \text{for all } h \in H_N^r. \qquad (5.5)$$

Here $\alpha > 0$ and the involved norms are defined as follows. For a function space X we abbreviate

$$L_\alpha^\infty X = L_\alpha^\infty((0, T); X) = \{ f : (0, T) \to X \mid \|f\|_{L_\alpha^\infty X} < \infty \}$$
$$\text{and} \quad \|f\|_{L_\alpha^\infty X} = \operatorname*{ess\,sup}_{t \in (0,T)} t^\alpha \|f(t)\|_X.$$

From known estimates on Green's functions we get corresponding bounds on $P(\mu)$ in terms of the simpler heat kernel.

Lemma 5.3 Estimate on fundamental solution, [12, Ch. 1, Sec. 8]). *For given dimension d, $r > 1 + \frac{d}{2}$, and diffusion coefficient σ^2 there exist a constant $C > 0$ and a continuous function $\hat{C} : [0, \infty) \to (0, \infty)$ such that*

$$|P(\mu)(t, x, x^0)| \leq \hat{C}(\|\mu\|_{H^r(\Omega; \mathbb{R}^d)}) t^{-\frac{d}{2}} \, \exp\left(-C \frac{|x-x^0|^2}{t} \right) =: G_0(t, x - x^0),$$
$$|\partial_{x^0} P(\mu)(t, x, x^0)| \leq \hat{C}(\|\mu\|_{H^r(\Omega; \mathbb{R}^d)}) t^{-\frac{d+1}{2}} \, \exp\left(-C \frac{|x-x^0|^2}{t} \right) =: G_1(t, x - x^0).$$

Notice that G_0 and G_1 are both time- and space-invariant under translation.

Remark 5.1 (Estimates and regularity of Green's function). *For the fundamental solution (instead of Green's function $P(\mu)$) the previous estimate can be found in [12, Ch. 1, Sec. 8]. For Green's functions $P(\mu)$ we find estimates on $\partial_x P(\mu)$ (instead* *of $\partial_{x^0} P(\mu)$) in [18, Ch. 4, Thm. 16.3]. However, those estimates can be transferred to estimates of $\partial_{x^0} P(\mu)$ by the fact that for general Green's functions g_μ it holds $g_\mu(t, x; \tau, x^0) = g_\mu^*(\tau, x^0; t, x)$, the latter being the Green function of the adjoint PDE. (Note, though, that the estimates in [18, Ch. 4, Thm. 16.3] are explicitly stated only for Dirichlet boundary conditions.) Similar results can also be found in [16, Sec. 4, Thm. 1].*

The estimates above imply

$$\|P(\mu)\|_{L_\alpha^\infty L^2(\Omega \times \Omega)}, \ \|\partial_{x_0} P(\mu)\|_{L_\beta^\infty L^2(\Omega \times \Omega)} \leq \tilde{C} \tag{5.6}$$

for some $\tilde{C} < \infty$ and $\alpha = \frac{d}{4}$, $\beta = \frac{d+2}{4}$, if μ is bounded in $H^r(\Omega; \mathbf{R}^d)$. We can see that with the following calculation

$$
\begin{aligned}
\|P(\mu)(t, \cdot, x^0)\|_{L^2(\Omega)}^2 &\leq \hat{C} t^{-d} \int_\Omega \exp\left(-2C \frac{|x - x^0|^2}{t}\right) dx \\
&\leq \hat{C} t^{-d} \int_{\mathbf{R}^d} \exp\left(-2C \frac{|x - x^0|^2}{t}\right) dx \\
&= \hat{C} t^{-d} \prod_{k=1}^d \int_{\mathbf{R}} \exp\left(-2C \frac{|x_k - x_k^0|^2}{t}\right) dx_k^i \\
&\simeq \hat{C} t^{-d} t^{\frac{d}{2}} \\
&\simeq \hat{C} t^{-\frac{d}{2}}.
\end{aligned}
$$

Since Ω is bounded the integration over the x^0 variable includes only another factor and thus we obtain

$$\|P(\mu)(t, \cdot, \cdot)\|_{L^2(\Omega \times \Omega)} \simeq \hat{C} t^{-\frac{d}{4}},$$

from which the claim follows. The computation for $\|\partial_{x_0} P(\mu)\|_{L_\beta^\infty L^2(\Omega \times \Omega)}$ is analogous. This tells us, for instance, that for $d = 1$, $P(\mu)$ is in $L^\infty L^2(\Omega \times \Omega)$ and thus for this special case we would not need to consider the time-weighted norms, but this is not true for $d > 1$.

Consider now the solution operator $\left(L_\mu + \frac{\partial}{\partial t}\right)^{-1}$ of the equation

$$\partial_t u + L_\mu u = f \qquad\qquad \text{in } [0, T] \times \Omega,$$
$$(\tfrac{\sigma^2}{2}\nabla u - \mu u)\cdot v = 0 \qquad\qquad \text{on } [0, T] \times \partial\Omega,$$
$$u = 0 \qquad\qquad \text{at } t = 0,$$

mapping the right hand side f to its solution u. It is linear since we assume trivial initial conditions. We will need a bound on this operator for the special case that the right hand side f is the divergence of a solution of (5.4) multiplied with a drift. Let us state this useful fact as a lemma regarding low dimensions.

Lemma 5.4 (Boundedness of solution operator). *Let $d \le 3$, $\mu \in H^r(\Omega; \mathbb{R}^d)$ for $r > \frac{d}{2} + 1$, and $g \in L_\alpha^\infty L^2(\Omega; \mathbb{R}^d) \cap L_\beta^\infty H^1(\Omega; \mathbb{R}^d)$ with $g \cdot v = 0$ on $\partial\Omega$. Let u solve $\partial_t u + L_\mu u = \nabla \cdot g$ in Ω with homogeneous initial and Neumann boundary conditions. There exists $\bar{C} > 0$ only depending on $\|\mu\|_{H^r(\Omega;\mathbb{R}^d)}$ and Ω such that $\|u\|_{L_0^\infty L^2(\Omega)} \le \bar{C}\|g\|_{L_\alpha^\infty L^2(\Omega)}.$*

Proof By Duhamel's principle, $u(t, x) = \int_0^t \int_\Omega P(\mu)(t - \tau, x, x^0)\nabla_{x^0} \cdot g(\tau, x^0)\,dx^0\,d\tau$ so that

$$\|u(t, \cdot)\|_{L^2(\Omega)} = \left\| \int_0^t \int_\Omega P(\mu)(t - \tau, \cdot, x^0)\nabla_{x^0} \cdot g(\tau, x^0)\,dx^0\,d\tau \right\|_{L^2(\Omega)}$$

$$= \left\| \int_0^t \int_\Omega \partial_{x^0} P(\mu)(t - \tau, \cdot, x^0) g(\tau, x^0)\,dx^0\,d\tau \right\|_{L^2(\Omega)}$$

$$\le \left\| \int_0^t \int_\Omega |G_1(t - \tau, \cdot - x^0)||g(\tau, x^0)|\,dx^0\,d\tau \right\|_{L^2(\Omega)}$$

$$= \left\| \int_0^t |G_1(t - \tau, \cdot)| * |g(\tau, \cdot)|\,d\tau \right\|_{L^2(\Omega)}$$

$$\le \int_0^t \||G_1(t - \tau, \cdot)| * |g(\tau, \cdot)|\|_{L^2(\Omega)}\,d\tau$$

$$\le \int_0^t \|G_1(t - \tau, \cdot)\|_{L^1(\mathbb{R}^d)} \|g(\tau, \cdot)\|_{L^2(\Omega;\mathbb{R}^d)}\,d\tau$$

$$\le \|g\|_{L_\alpha^\infty L^2(\Omega;\mathbb{R}^d)} \int_0^T \tau^{-\alpha} \|G_1(t - \tau, \cdot)\|_{L^1(\mathbb{R}^d)}\,d\tau$$

$$= \hat{C}(\|\mu\|_{H^r(\Omega;\mathbb{R}^d)}) \left(\frac{\pi}{C}\right)^{d/2} \|g\|_{L_\alpha^\infty L^2(\Omega;\mathbb{R}^d)} \int_0^t \tau^{-\alpha}\frac{1}{\sqrt{t - \tau}}\,d\tau$$

$$= \tilde{C}\|g\|_{L_\alpha^\infty L^2(\Omega;\mathbb{R}^d)}, \qquad\qquad (5.7)$$

where we used integration by parts, lemma 5.3, the translation invariance of G_1, Young's convolution inequality, and $\alpha = \frac{d}{4} < 1$. $\square$

The next step is to prove the continuity of operator P in the sense that for $\|h\|_\infty \to 0$ it holds $\|P(\mu + h) - P(\mu)\|_{L_\alpha^\infty L^2(\Omega \times \Omega)} \to 0$.

Lemma 5.5 (Continuity of Green function operator). *Let $d \leq 3$ and $\mu \in H_N^r$ for $r > \frac{d}{2} + 1$. There exists $\tilde{C} > 0$ depending only on $\|\mu\|_{H^r(\Omega;\mathbb{R}^d)}$ such that*

$$(1 - \tilde{C}\|h\|_{L^\infty})\|P(\mu + h) - P(\mu)\|_{L_\alpha^\infty L^2(\Omega \times \Omega)}$$

$$\leq \tilde{C}\|h\|_{L^\infty}\|P(\mu)\|_{L_\alpha^\infty L^2(\Omega \times \Omega)} \qquad \text{for all } h \in H_N^r.$$

Proof For fixed x^0 define $\tilde{u}_\mu^h(t, x) := [P(\mu + h) - P(\mu)](t, x, x^0)$ and $u_h(t, x) := P(\mu + h)(t, x, x^0)$, then $\tilde{u}_\mu^h$ satisfies

$$\partial_t \tilde{u}_\mu^h + L_\mu \tilde{u}_\mu^h u = -\nabla \cdot (h u_h) \qquad \text{in } [0, T] \times \Omega,$$

$$\left(\tfrac{\sigma^2}{2}\nabla u - \mu u\right) \cdot v = 0 \qquad \text{on } [0, T] \times \partial\Omega,$$

$$u = 0 \qquad \text{at } t = 0,$$

Due to $u_h \in L_\alpha^\infty L^2(\Omega) \cap L_\beta^\infty H^1(\Omega)$ by (5.6) and $r > 1 + \frac{d}{2}$ we can apply lemma 5.4 with $g = -h u_h$ to obtain

$$\|\tilde{u}_\mu^h\|_{L^2(\Omega)} \leq \bar{C}\|h u_h\|_{L_\alpha^\infty L^2(\Omega)} \leq \bar{C}\|h\|_{L^\infty(\Omega)}\|u_h\|_{L_\alpha^\infty L^2(\Omega)},$$

which implies

$$\|\tilde{u}_\mu^h\|_{L_\alpha^\infty L^2(\Omega)} \leq \tilde{C}\|h\|_{L^\infty(\Omega)}\|u_h\|_{L_\alpha^\infty L^2(\Omega)}$$

for $\tilde{C} = T^\alpha \bar{C}$. Uing the triangle inequality we obtain

$$\|P(\mu + h) - P(\mu)\|_{L_\alpha^\infty L^2(\Omega \times \Omega)} \leq \tilde{C}\|h\|_{L^\infty(\Omega)}\|P(\mu + h)\|_{L_\alpha^\infty L^2(\Omega \times \Omega)}$$

$$\leq \tilde{C}\|h\|_{L^\infty(\Omega)}[\|P(\mu)\|_{L_\alpha^\infty L^2(\Omega \times \Omega)} + \|P(\mu + h) - P(\mu)\|_{L_\alpha^\infty L^2(\Omega \times \Omega)}]$$

and thus the desired inequality. $\square$

Now we consider differentiability of operator P. to characterize the Fréchet-derivative of the operator one can compute the directional derivative of equation

(5.4) at position μ in direction h for some $h \in H_N^r$. This yields a PDE that indeed defines the Fréchet derivative of operator P. With this characterization of the derivative we are able to show the disired weak tangential cone condition.

Proposition 5.1 (Differentiability and tangential cone condition for Green function operator). *Let $d \leq 3$ and $\mu \in H_N^r$ for $r > \frac{d}{2} + 1$. Define the operator $P'(\mu) : H_N^r \to L_\alpha^\infty L^2(\Omega \times \Omega) \cap L_\beta^\infty H^1(\Omega \times \Omega)$ via*

$$P'(\mu)[h](t, x, x^0) = u_\mu^h(t, x),$$

where u_μ^h is the solution of

$$\partial_t u + L_\mu u = -\mathrm{div}(h\, P(\mu)(\cdot, \cdot, x^0)) \qquad \text{in } [0, T] \times \Omega,$$
$$(\tfrac{\sigma^2}{2}\nabla u - \mu u) \cdot \nu = 0 \qquad \text{on } [0, T] \times \partial\Omega,$$
$$u = 0 \qquad \text{at } t = 0,$$

then the tangential cone condition (5.5) holds for $\mu^\dagger = \mu$.

Proof For fixed x^0 let again $\tilde{u}_\mu^h(t, x) = [P(\mu + h) - P(\mu)](t, x, x^0)$, then $\tilde{u}(t, x) = [P(\mu + h) - P(\mu) - P'(\mu)[h]](t, x, x^0)$ satisfies

$$\partial_t \tilde{u} + L_\mu \tilde{u} = -\mathrm{div}(h\tilde{u}_\mu^h)$$

with homogeneous initial and Neumann boundary conditions. Again appealing to lemma 5.4 (which is possible due to $\tilde{u}_\mu^h \in L_\alpha^\infty L^2(\Omega) \cap L_\beta^\infty H^1(\Omega)$ by (5.6) and $r > 1 + \frac{d}{2}$) we obtain

$$\|\tilde{u}\|_{L^2(\Omega)} \leq \bar{C}\|h\tilde{u}_\mu^h\|_{L_\alpha^\infty L^2(\Omega)} \leq \bar{C}\|h\|_{L^\infty}\|\tilde{u}_\mu^h\|_{L_\alpha^\infty L^2(\Omega)}$$

and thus

$$\|\tilde{u}\|_{L_\alpha^\infty L^2(\Omega)} \leq \bar{C}T^\alpha\|h\|_{L^\infty}\|\tilde{u}_\mu^h\|_{L_\alpha^\infty L^2(\Omega)}.$$

Finally squaring the inequality and integrating over the variable x^0 yields the result.

Remark 5.2 *Notice that the right-hand side of (5.5) is $o(\|h\|_{L^\infty(\Omega)})$ due to lemma 5.5, which means that P' is indeed the Fréchet-derivative of P in the L^∞-topology.*

References

1. Hans Wilhelm Alt. *Linear functional analysis*. Universitext. Springer-Verlag London, Ltd., London, german edition, 2016. An application-oriented introduction.
2. Radu Ioan Boţ and Bernd Hofmann. An extension of the variational inequality approach for obtaining convergence rates in regularization of nonlinear ill-posed problems. *J. Integral Equations Appl.*, 22(3):369–392, 2010.
3. J. M. Borwein and A. S. Lewis. Convergence of best entropy estimates. *SIAM J. Optim.*, 1(2):191–205, 1991.
4. Constantin Băcuţă, Anna L. Mazzucato, Victor Nistor, and Ludmil Zikatanov. Interface and mixed boundary value problems on n-dimensional polyhedral domains. *Doc. Math.*, 15:687–745, 2010.
5. Martin Burger and Stanley Osher. Convergence rates of convex variational regularization. *Inverse Problems*, 20(5):1411–1421, 2004.
6. Gregory S. Chirikjian. *Stochastic models, information theory, and Lie groups. Vol. 1.* Applied and Numerical Harmonic Analysis. Birkhäuser Boston, Ltd., Boston, MA, 2009. Classical results and geometric methods.
7. Fabian Dunker and Thorsten Hohage. On parameter identification in stochastic differential equations by penalized maximum likelihood. *Inverse Problems*, 30(9):095001, 20, 2014.
8. Heinz W. Engl, Martin Hanke, and Andreas Neubauer. *Regularization of inverse problems*, volume 375 of *Mathematics and its Applications*. Kluwer Academic Publishers Group, Dordrecht, 1996.
9. Heinz W. Engl, Karl Kunisch, and Andreas Neubauer. Convergence rates for Tikhonov regularisation of nonlinear ill-posed problems. *Inverse Problems*, 5(4):523–540, 1989.
10. Jens Flemming. *Generalized Tikhonov regularization and modern convergence rate theory in Banach spaces*. Berichte aus der Mathematik. Shaker Verlag, Aachen, 2012.
11. Jens Flemming and Bernd Hofmann. Convergence rates in constrained Tikhonov regularization: equivalence of projected source conditions and variational inequalities. *Inverse Problems*, 27(8):085001, 11, 2011.
12. Avner Friedman. *Partial differential equations of parabolic type*. Prentice-Hall, Inc., Englewood Cliffs, NJ, 1964.

13. David Gilbarg and Neil S. Trudinger. *Elliptic partial differential equations of second order*. Classics in Mathematics. Springer-Verlag, Berlin, 2001. Reprint of the 1998 edition.

14. B. Hofmann, B. Kaltenbacher, C. Pöschl, and O. Scherzer. A convergence rates result for Tikhonov regularization in Banach spaces with non-smooth operators. *Inverse Problems*, 23(3):987–1010, 2007.

15. Bernd Hofmann and Masahiro Yamamoto. On the interplay of source conditions and variational inequalities for nonlinear ill-posed problems. *Appl. Anal.*, 89(11):1705–1727, 2010.

16. A. M. Il'in, A. S. Kalashnikov, and O. A. Oleinik. Second-order linear equations of parabolic type. *Tr. Semin. im. I. G. Petrovskogo*, (21):9–193, 341, 2001.

17. Hui-Hsiung Kuo. *Introduction to stochastic integration*. Universitext. Springer, New York, 2006.

18. O. A. Ladyženskaja, V. A. Solonnikov, and N. N. Ural' ceva. *Lineĭ nye i kvazilineĭ nye uravneniya parabolicheskogo tipa*. Izdat. "Nauka", Moscow, 1967.

19. Pascal Massart. About the constants in Talagrand's concentration inequalities for empirical processes. *Ann. Probab.*, 28(2):863–884, 2000.

20. Peter Mathé and Bernd Hofmann. How general are general source conditions? *Inverse Problems*, 24(1):015009, 5, 2008.

21. V. Maz'ya and T. Shaposhnikova. An elementary proof of the Brezis and Mironescu theorem on the composition operator in fractional Sobolev spaces. *J. Evol. Equ.*, 2(1):113–125, 2002.

22. Christiane Pöschl. *Tikhonov Regularization with General Residual Term*. PhD thesis, 10 2008.

23. Elena Resmerita and Otmar Scherzer. Error estimates for non-quadratic regularization and the relation to enhancement. *Inverse Problems*, 22(3):801–814, 2006.

24. Thomas I. Seidman and Curtis R. Vogel. Well-posedness and convergence of some regularisation methods for nonlinear ill posed problems. *Inverse Problems*, 5(2):227–238, 1989.

25. Daniel Tenbrinck, Nikolas Uesseler, Philipp Wacker, and Benedikt Wirth. On map estimates and source conditions for drift identification in sdes, 2025. arXiv:2507.18443 [math.NA].

26. Frank Werner. *Inverse Problems with Poisson Data: Tikhonov-Type Regularization and Iteratively Regularized Newton Methods*. PhD thesis, University of Göttingen, Göttingen, Germany, 2012. Doctoral dissertation.

27. Frank Werner and Thorsten Hohage. Convergence rates in expectation for Tikhonov-type regularization of inverse problems with Poisson data. *Inverse Problems*, 28(10):104004, 15, 2012.